Springer Series on Polymer and Composite Materials

Series editor

Susheel Kalia, Dehradun, India

More information about this series at http://www.springer.com/series/13173

Abhijit Bandyopadhyay · Srijoni Sengupta
Tamalika Das

Hyperbranched Polymers
for Biomedical Applications

Springer

Abhijit Bandyopadhyay
Department of Polymer Science and
  Technology
University of Calcutta
Kolkata, West Bengal
India

Tamalika Das
Department of Polymer Science and
  Technology
University of Calcutta
Kolkata, West Bengal
India

Srijoni Sengupta
Department of Polymer Science and
  Technology
University of Calcutta
Kolkata, West Bengal
India

ISSN 2364-1878   ISSN 2364-1886 (electronic)
Springer Series on Polymer and Composite Materials
ISBN 978-981-13-4895-2   ISBN 978-981-10-6514-9 (eBook)
https://doi.org/10.1007/978-981-10-6514-9

This Springer imprint is published by Springer Nature
The registered company is Springer Nature Singapore Pte Ltd.
The registered company address is: 152 Beach Road, #21-01/04 Gateway East, Singapore 189721, Singapore

Contents

1 Introduction.. 1
  1.1 Introduction to the World of Polymers...................... 1
  1.2 Conventional Polymers with Special Reference
      to Architectural Polymers................................. 3
  1.3 Dendrimers: Advantages and Disadvantages.................. 6
  1.4 Introduction to Hyperbranched Polymers.................... 9
  1.5 Conclusion... 13
  References... 13

2 Part I—Synthesis of Hyperbranched Polymers: Step-Growth
  Methods.. 15
  2.1 Introduction to Theoretical Approaches in Hyperbranched
      Polymerization... 15
  2.2 Hyperbranched Polymers from AB$_x$-Type Monomers.......... 16
      2.2.1 Carbon–Carbon Coupling Reactions................... 19
      2.2.2 Carbon-Hetero Atom Coupling Reactions.............. 35
  2.3 Hyperbranched Polymers from A$_2$ + B$_3$ Monomer Pairs
      and Other Couple Monomer Methodologies................... 49
  2.4 Drawbacks of Hyperbranched Polymerization Techniques
      and Possible Remedies.................................... 52
  2.5 Conclusion... 57
  References... 57

3 Part II—Synthesis of Hyperbranched Polymers: Mixed
  Chain-Growth and Step-Growth Methods......................... 65
  3.1 Introduction to Simultaneous Step- and Chain-Growth
      Methodologies.. 66
  3.2 Radical Polymerization................................... 67
  3.3 Proton Transfer Polymerization........................... 72
  3.4 Self-Condensing Vinyl Polymerization/Copolymerization,
      Self-Condensing Ring Opening Polymerization
      and Controlled/Living Polymerization..................... 76

3.5   Hypergrafting   97
    3.5.1   Homogeneous Grafting—Hyperbranched-Graft-
         Hyperbranched Copolymers   97
    3.5.2   Homogeneous Grafting—Linear-Graft-
         Hyperbranched Copolymers   98
    3.5.3   Homogeneous Grafting—Linear-Block-
         Hyperbranched Copolymers   99
    3.5.4   Heterogeneous Grafting............................   101
    3.5.5   Hypergrafting onto Living Cells   102
3.6   Conclusion   102
References.   102

4   Structure–Property Relationship of Hyperbranched Polymers   109
4.1   Introduction to Intrinsic Properties of Hyperbranched
    Polymers.   109
4.2   Degree of Branching.   110
    4.2.1   Determination of DB.   110
    4.2.2   Methods to Determine DB   111
4.3   Influence of the Branching Architecture
    on the End Properties.   115
4.4   Solution Properties with Special Reference
    to Hyperbranched Architecture   116
4.5   Molecular Mass and Molar Mass Distribution   118
4.6   Bulk Properties.   120
    4.6.1   Thermal Properties   120
    4.6.2   Rheological Properties.   123
4.7   Special Properties Related to Latest Biological Applications   126
    4.7.1   Biodegradability and Biocompatibility
         of Hyperbranched Polymers   126
    4.7.2   Self-Assembly of the Hyperbranched Polymers.   129
    4.7.3   Encapsulation by Hyperbranched Polymers.   129
4.8   Conclusion   132
References.   132

5   Latest Biomedical Applications of Hyperbranched
Polymers: Part 1: As Delivery Vehicle.   135
5.1   Introduction to the Concept of Targeted Delivery.   135
5.2   Encapsulation Ability of Hyperbranched Polymers.   136
5.3   Hyperbranched Polymers in Controlled Drug Delivery.   138
5.4   Hyperbranched Polymers in Protein Delivery   141
5.5   Hyperbranched Polymers in Gene Delivery   143
5.6   Conclusion   149
References.   149

6   Part II: In Bioimaging 151
   6.1   Introduction to Diagnosis via Bioimaging.................... 151
   6.2   Hyperbranched Polymers as Fluorescent Probes.............. 152
   6.3   Hyperbranched Polymers as MRI Contrast Agents............ 156
   6.4   Hyperbranched Polymers in Nuclear Tomographic Imaging...... 158
   6.5   Hyperbranched Polymers for Multimodal Imaging............. 160
   6.6   Conclusion 163
   References... 163

7   Part III: Tissue Engineering............................... 165
   7.1   Introduction 165
   7.2   Hyperbranched Polymers as Tissue Scaffold Component 165
   7.3   Hyperbranched Polymers as Cell and Tissue Adhesives......... 172
   7.4   Conclusion 175
   References... 175

8   Conclusion 177

About the Authors

Dr. Abhijit Bandyopadhyay holds an M.Tech. and Ph.D. in Polymer Science and Technology and is currently Head of the Department of Polymer Science and Technology, University of Calcutta. He is also a Technical Director at the South Asian Rubber and Polymers Park (SARPOL) in West Bengal, India. He is a former Assistant Professor of the Rubber Technology Centre at the Indian Institute of Technology Kharagpur (IIT Kharagpur), India. He has more than 10 years of teaching and research experience and has published more than 75 research papers in high-impact international journals, four book chapters, one book, and holds one Indian patent. He has received numerous awards, including the Young Scientist Award from the Materials Research Society of India, Calcutta Chapter; the Young Scientist Award from the Department of Science and Technology, Government of India; and the Career Award for Young Teachers from the All India Council for Technical Education, Government of India. He is a life member of the Society for Polymer Science, India, Associate Life Member of the Indian Institute of Chemical Engineers, and a Fellow of the International Congress of Environmental Research. He also serves on the Editorial Boards of various international journals.

Ms. Srijoni Sengupta is a research scholar at the Department of Polymer Science and Technology, University of Calcutta, India. She received her initial degree in Chemistry (Hons) from Lady Brabourne College (Kolkata) in 2010 before completing her B.Tech. (2013) and M.Tech. (2015) in Polymer Science and Technology at the University of Calcutta. She is currently pursuing her Ph.D. on the "Synthesis, Study of Structure-Property Relationships and Potential Applications of Hyperbranched Polymers via Polycondensation Technique." She has published a number of papers in research journals and is presently engaged in an INSPIRE Fellowship from the Department of Science and Technology, Government of India.

Ms. Tamalika Das graduated in Chemistry (Hons) from Scottish Church College (Kolkata) in 2008. She subsequently completed her B.Tech. (2011) and M.Tech. (2013) degrees at the Department of Polymer Science and Technology at the

University of Calcutta. She was awarded the gold medal from the University of Calcutta twice (during both her B.Tech. and M.Tech.). She did her M.Tech. project at the Indian Association for the Cultivation of Science (Kolkata). Currently, she is pursuing her doctoral research at the Department of Polymer Science and Technology at the University of Calcutta. Her main area of interest is hyper-branched polymers. She has published three papers in high-impact international journals and is currently engaged in an INSPIRE Fellowship from the Department of Science and Technology, Government of India.

Abstract

This book presents a detailed study on a new class of branched polymers, hyperbranched polymers. The synthesis strategies of these special classes of polymers are discussed elaborately. Polymers have become a basic necessity in human life due to its wide range of properties. The conventional polymers have three basic structural alteration—linear, branched, and fully crosslinked form. Each of the forms has respective application area with its individual merits and demerits. Later polymer scientists tried to explore more with the properties of the polymers by altering its shape and structure. In order to develop different kinds of architectural polymers, dendrimers came into existence. Dendrimers have symmetrically branched structure with high availability of tunable terminal functional groups. Hyperbranched polymers are a subclass of dendrimers with unsymmetrical random branching. The unusual structure, specific tunable properties, and wide scope of applications of the hyperbranched polymers have triggered the interest in polymer scientists and technologists all over the world. The advantage of the hyperbranched polymers over the dendrimers is its ease of synthesis. The one-pot synthesis technique of the hyperbranched polymers makes it more feasible and acceptable for large-scale production. The structure–property relationship of hyperbranched polymers are also discussed, which makes it superior compared to its linear analogue. The alterable functional groups present at the terminal ends of the branches makes HBP a good option in the biomedical field. Biocompatible and biodegradable HBPs are the recent interests for the polymer technologists as its potential in biomedical area is huge. In fact in the past few years, the application of HBPs in biological world has experienced a rapid growth, which is discussed in details in this book.

This book will be a hope to young researchers developing modified architectural polymers in an attempt to open up newer structure and properties. As here we have specifically elaborated suitable characteristic properties of each of the probable biological applications of the HBPs, the book can be a guidance to scientists and technologists who can commercially try these newly developed techniques for faster and better treatment so that these new medical field is explored more.

Chapter 1
Introduction

1.1 Introduction to the World of Polymers

Polymers were existent in nature since the beginning of life. It existed in the form of DNA, RNA, proteins, and polysaccharides and played a pivotal role in the evolution of the biological world. Cellulose and starch were the most common naturally occurring polymers, both having glucose as the monomeric unit conjoined through the glycosidic linkage [1]. The next most common natural polymer was natural rubber, extracted from the latex of *Hevea Brasiliensis* (rubber tree) [2]. However, during the 1800s the growing demand for new materials initiated the chemical modifications of the natural polymers to develop semi-synthetic polymers. Vulcanization of rubber, invented by Charles Goodyear in the nineteenth century [3], so far has been the best example of such modification. In the early twentieth century, synthetic polymers came into existence with the synthesis of Bakelite by Leo Baekeland [4]. He developed the said resin from synthetic resin with high hardness. It was soon followed by the production of Rayon, a synthetic fiber from cellulose. Following these, a series of important synthetic polymers like nylon [5], styrene butadiene rubber, acrylic polymers, neoprene rubber, and many others were synthesized under commercial scale during the period of World War II to meet the increasing demand [6].

Herman Staudinger has been regarded as the pioneer for originating a systematic study in polymer science. He provided a new definition of polymer and with his skill and exuberance, a wide research area got unveiled before the scientific world further exploration. In 1953, Staudinger was awarded the Nobel Prize for his contribution toward the field of high polymer chemistry [7].

Invention of semi-synthetic and later the synthetic polymers appended a new class of viscoelastic material in the field of material science and technology. Starting from daily use to world's most sophisticated technology, the synthetic polymers were invaded in every possible corner of applications pertaining to unique molecular conformation and chemistry of the macromolecules that eventually

© Springer Nature Singapore Pte Ltd. 2018
A. Bandyopadhyay et al., *Hyperbranched Polymers for Biomedical Applications*,
Springer Series on Polymer and Composite Materials,
https://doi.org/10.1007/978-981-10-6514-9_1

renders a unique combination of physicochemical and mechanical properties. Polymer industry so far is the fastest growing industry and almost everyday, the world sees an invention in polymer technology that comes into the limelight.

Apart from origin, the polymers are found in different structures and shapes. Polymers are available in three basic structures—linear, branched, and crosslink or network. Figure 1.1 shows a demonstration of these basic structures. Linear form is the simplest of the three but most difficult to synthesize. Stringent conditions are required during the synthesis to avoid slightest of the branch formation. Conversely, branched structures are easier to synthesize as it is entropically more favorable. Monomer(s) having average functionality two would result into a linear polymeric structure, while an average functionality more than two results into the branched form. The degree of branching that occurs during polymerization can be influenced by the functionality of the monomers [8]. For example, in a free radical polymerization of styrene, addition of divinylbenzene, which has a functionality of two, will result in the formation of a branched polymer. Branches can form when the growing end of a polymer molecule reaches either (a) back around onto itself or (b) onto another polymer chain; in either of these cases, a mid-chain growth side is created via an abstraction of a hydrogen atom [9]. Crosslink or network formation is an extended form of branching. A network structure is formed through an extensive branching reaction and a three-dimensional net-type structure is created at the end. However, each type of polymers has their independent shape and size following intra- as well as inter-molecular forces of interaction and external stimulation (solvent, temperature, stress, etc.).

Polymers are said to be branched when the linear chains diverge in some way. Branching more often can result during the synthesis or in some cases from the post-synthesis modification of the polymer. Branching leads to more compact and dense morphology compared to the linear analogues, and generates a radically different melt and flow behavior [10]. In fact, there has been much interest paid recently toward the branched polymers and beyond, as a unique method of controlling the end properties even for the well-known polymers. In addition, polymers may intrinsically have a wide variety of branched structures depending on how they

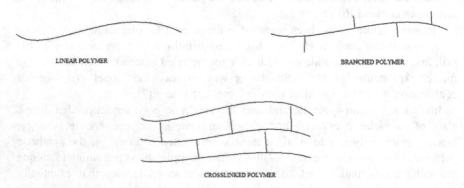

Fig. 1.1 Pictorial views of linear, branched, and crosslinked polymer structures

have been synthesized or modified. The branching can further be characterized by the average length of the branches, i.e., either the branches were long or short and indeed their distribution. Long-chain branches affect the size and density of polymer molecules and are easier to measure by GPC than short-chain branches [11]. In general, a higher degree of branching imbibes compactness in the polymer structure. Branching also affects the chain entanglement density and the ability of chains to slide past one another, under stress and thus affects the bulk physical properties. However, long-chain branches may increase the polymer strength, toughness, and the glass transition temperature (T_g) due to an increasing number of entanglements per chain. A random and short-chain branch, on the other hand, may reduce polymer strength due to disruption of the ability of the chains to interact with each other or crystallize [12].

A classic example of the effect of branching on physical properties can be found in the case of polyethylene. High-density polyethylene or HDPE historically has a very low degree of branching, is relatively stiff, and is used in high-strength applications such as bullet-proof vests. Low-density polyethylene (LDPE), on the other hand, has significant numbers of both long- and short-chain branches and is relatively flexible, and used in relatively soft grade applications such as plastic films [13].

1.2 Conventional Polymers with Special Reference to Architectural Polymers

With the invention of living polymerization, the synthesis of "tailor-made" polymers with specific architectures had become easier and adaptable. Synthesis of coveted architectures such as alternate, block, star, comb, brush, dendritic, and ring (co)polymers was made possible using the living polymerization technique. A schematic in Fig. 1.2 demonstrates the major architectures developed so far. The following section, very briefly, describes each of these polymers.

- **Alternating, Random, Block, and Graft Copolymers**:
  When a polymer has one type of monomer unit linkage, it is known as a homopolymer and when two or more different types of monomers are linked it is known as a copolymer. **Alternating copolymers** are generated when the monomer units are linked in regular alternating fashion. Random linkage of the monomers leads to **random copolymer** structure. **Block copolymers** consist of two or more homopolymer unit(s) simply linked by a covalent bond. **Graft copolymers** are a differently branched copolymer where the composition of the main chain is different from the side chain; nevertheless, the individual chains may be either homopolymer or copolymer themselves [14].
- **Star Polymer**:
  Star-shaped polymers are the simplest class of branched polymers with a structure consisting of few linear chains connected to a central core. Star-shaped polymers in which the side linear arms have equivalent length and structure are

(a)

(b)

(c)

(d)

(e)

(f)

(g)

(h)

(i)

(j)

(k)

(l)

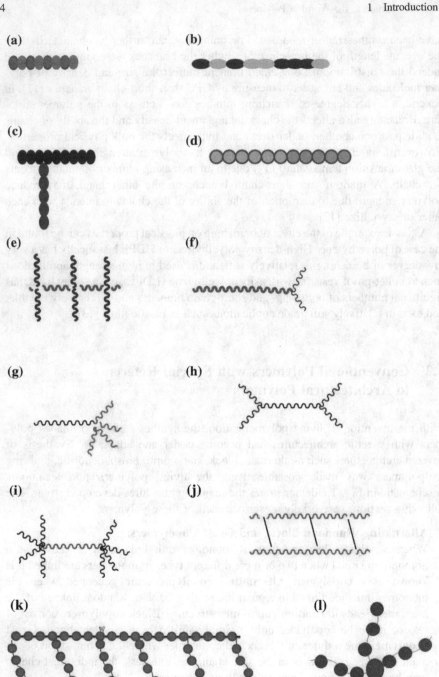

◀**Fig. 1.2** Different architectural polymers: **a** Alternating copolymer, **b** Random copolymer, **c** Graft copolymer, **d** Block copolymer, **e** Brush copolymer, **f** AB_2 star polymer, **g** Palm tree AB_n, **h** H-shaped B_2AB_2, **i** Dumbelled shaped, **j** Ladder-shaped polymer, **k** Comb polymer, **l** Star polymer

considered homogeneous, and ones with variable lengths and structures are considered heterogeneous. Star-shaped polymers have unique properties, such as compact structure, high arm density, efficient synthetic routes, and unique rheological properties [15].

- **Polymer Brush**:
  Polymer brush is a series of linear polymer chains attached to one end to a surface. The brushes are often characterized by the high density of grafted chains [16].
- **Comb Polymer**:
  A comb polymer molecule consists of a main chain with many branch points and linear side chains. When the side arms are identical, the comb polymer molecule is said to be regular [17].
- **Ladder Polymer**:
  The ladder polymers, as suggested by the name, have a similarity between the schematic projection of the plane of the macromolecule and a ladder. It is synthesized either by cyclization of the corresponding linear polymers or by direct polymerization of monomers [18].

Degree of branching is an important parameter to describe intricacy of a branched polymer which is defined by the ratio of branched, terminal, and linear units in the polymer architecture [19].

Branching index, another important terminology used to define the extent of branching alike degree of branching, measures the effect of long-chain branches on the size of a macromolecule in solution. The branching index, "g", is expressed through an expression reported as Eq. 1.1:

$$g = S_b^2 / S_1^2, \tag{1.1}$$

where S_b is the mean square radius of gyration of the branched macromolecule in a given solvent and S_1 is the mean square radius of gyration of the linear analogue. A value greater than 1 indicates an optimistic rise in radius of gyration due to branching [20].

To avoid gelation during synthesis of architectural polymers, either stoichiometric imbalance or use of a chain transfer agent is required as a tool. Thiols proved to be a very effective chain transfer agent in various forms of reactions.

1.3   Dendrimers: Advantages and Disadvantages

It has been a great challenge to the polymer scientists and technologists to develop special architectural polymers with tunable functionalities in order to meet the faster growing demand for versatile applications. Thus, an interdependence of shape on structure has been an extremely important research subject nowadays in polymer science and technology. Historically, the first breakthrough in architectural polymers was made by Doe Chemicals & Co. in the year 1984 through the synthesis of "Dendritic polymers" or dendrimers [21]. The dendrimer was synthesized from polyamidoamine (PAMAM) which appeared like a geometrically symmetrical branched tree. The name dendrimer was coined from the biological entity, Dendron, which has a very similar shape. Dendrimers have tunable functional end groups which made it applicable for a wide range of applications. It had evenly distributed symmetrical branches with enhanced physical and chemical properties. The degree of branching in the dendrimers determined its generation, such as first generation, second generation, etc. The synthesis of different grades of dendrimers became a prime important area for the polymer scientists. Dendrimers are synthesized both by convergent and divergent methodologies [22].

In **Divergent Methodology**, outward growth occurs from a multifunctional central core molecule [23]. The core molecule reacts with monomer molecules containing one active site and two dormant groups giving the first-generation dendrimer. Then the new terminal dormant functional group of the molecule is activated for further reactions with more monomers and the process is repeated as a dendrimer is built layer after layer giving multiple generations (Fig. 1.3). Although the divergent approach was effective for the production of dendrimers on larger scale, there are few disadvantages in this process. Structural defects in the dendrimers occur due to the side reactions and incomplete reactions of the end groups. Again to prevent side reactions and to force reactions to completion, large excess of reagents is required which create difficulties in the purification of the final product.

To prevent the disadvantages of the divergent synthesis, the **convergent method** was developed [24]. In the convergent approach, the dendrimer was developed starting from the end groups toward the inward core. As the growing branched polymeric arms, called dendrons, were large enough, they further got attached to a multifunctional core molecule (Fig. 1.4). The convergent growth method has many advantages such as it is relatively easy to purify the desired product and the occurrence of defects in the final structure is minimized, unlike the divergent method. It becomes easier to alter the properties of the dendrimer by just simply tuning the functional groups at the terminal end of the macromolecule through this method. But the main disadvantage of the convergent approach is the formation of higher generations that could be restricted owing to the steric hindrance created due to interactions between the dendrons and the core molecule.

Unlike linear polymers, dendrimers are monodisperse. The classical routes to linear polymerization eventually ended up with a spectrum of sizes of molecules, whereas the size and molecular mass of dendrimers can be specifically controlled

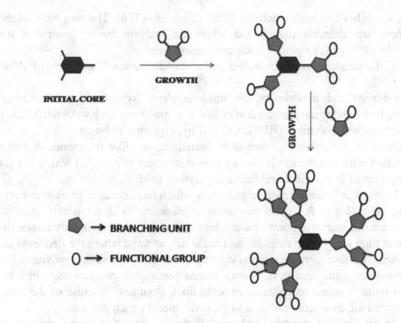

Fig. 1.3 Divergent synthesis route for dendrimers emerging from an initial core

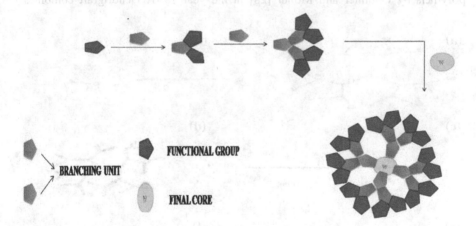

Fig. 1.4 Convergent synthesis route for dendrimers

during synthesis. Because of their unique molecular architecture, dendrimers show some significantly improved physical and chemical properties when compared to its linear analogues. In solution, linear chains exist as flexible coils, whereas in dendrimers it forms a tightly packed globular structure which greatly influences its rheological properties. Dendrimer solutions have comprehensively lower viscosity than the linear polymers [25]. The presence of "too" many chain ends is responsible for high solubility and high reactivity in the dendrimers [26]. The properties of the dendrimers are alterable due to the adjustable end functional groups of the macromolecules, which opens a wide application field.

By far, the dendrimers are classified under six subclasses. These are as follows [27]:

(i) Dendrons and dendrimers, (ii) linear-dendritic hybrids, (iii) dendronized polymers, (iv) dendrigrafts or dendrimer-like star macromolecules (DendriMacro), (v) hyperbranched polymers (HBP), and (vi) hypergrafted polymers.

Dendrons or dendrimers are regularly branched, tree-like fragments. A linear chain grafted with dendrimers leads to a hybrid structure (Fig. 1.5b) which has got wide importance in complex architectural polymer field.

Dendronized polymers are linear polymers which have dendrons linked at every repeating unit (Fig. 1.5c). The two main approaches to develop this class of polymers are the macromonomer route where a monomer which already carries the dendron of final size is polymerized, and the attach-to route where the dendrons are constructed by subsequent generation directly on an already existing polymer. The macromonomer route results in shorter chains for higher generations, while the attach-to route is prone to generate imperfections genuinely because of the enormous number of chemical reactions to be performed in each stage.

The fourth class of dendritic polymers is the dendrigraft systems, introduced simultaneously as comb-burst polymers by Tomalia et al. [21] and as arborescent polymers by Gauthier and Möller [28] in the year 1991. Dendrigraft combines

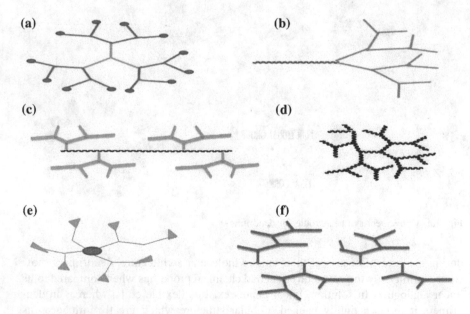

Fig. 1.5 Classification of dendrimers: **a** Dendrimers, **b** linear-dendritic hybrids, **c** dendronized polymers, **d** dendrigrafts or dendrimer-like star macromolecules (DendriMacro), **e** hyperbranched polymers (HBP), **f** hypergrafted polymer

features of dendrimers and hyperbranched polymers. Synthesis of a dendrigraft follows a generation-based growth scheme similar to dendrimers but uses polymeric chains as the initial building block. Thus, a very rapid increase in molecular weight per generation occurs in the former and a high molecular weight branched polymers are produced within a few steps. Since the grafting reaction is a random process, the branched chains are hyperbranched in nature. Even though the architecture of dendrigraft polymers is not as symmetrical as the dendrimers, the molecular weight distribution achieved for these class of materials lies within a narrow region [29].

Besides the interest in ordinary hyperbranched homopolymers or hb, grafted hbs have been increasingly used for the generation of hybrid structures, be it in the field of complex polymer topologies or for surface and particle functionalization [30].

While the first four subclasses have ideal structures with the degree of branching (DB) of nearly 1.0, the hyperbranched polymers exhibit random and irregular branching with much lesser degree of branching.

1.4   Introduction to Hyperbranched Polymers

The principal advantages of dendrimers include better solubility, lower solution/melt viscosity (which ensures easy and high rate of processing), high degree of functionality, high segment density, high thermal stability, etc. as compared to the conventionally structured polymers [31]. But the main disadvantage with dendrimers and other symmetrical architectural polymers is the inability for large-scale production mainly due to complicated multistep synthetic procedure and much higher production cost.

So the scientists, worldwide, tried to develop an alternative macromolecule as an effective substitute for dendrimers. The term "Hyperbranched polymers" or hb thus thrived in the scientific community as a "real and effective" substitute to high-cost dendrimers. The name "hyperbranched" was coined by the researchers of DuPont, Kim, and Webster in the early 1980s to define dendritic polymers having unsymmetrical branching [32]. They are highly branched and have three-dimensional globular structures. Hbs were first produced during late 1980s by one-step polycondensation of AB_2-type monomer. After the production of the first international patented hb, polyphenylene, curiosity toward modification of architectural polymers enhanced tremendously [33]. Aware of the fact that dendrimers were too hard to produce on commercial scale, researchers were keenly interested to develop hb polymers on a commercially viable scale. They tried to develop an architecture in which the branching is denser and the total number of functional groups is also higher. In fact, the history of hb polymers is quite old. Long ago, Berzelius synthesized a resin by reacting tartaric acid with glycerol ($A_2B_2 + B_3$) [34]. The concept of hyperbranching at that time was not prevalent, however, due to the higher functionality of the monomers the prepared resin was highly branched in structure. During 1901, several other scientists like Watson Smith, Callahan, Arsem, Dawson, and Howell investigated further with a bifunctional monomer phthalic anhydride monomer and a trifunctional monomer glycerol [35]. Though Bakelite, the first commercial synthetic plastic produced in 1909, was a phenolic thermoset, the precursor was a randomly branched semi-polymer because of the bifuntional formaldehyde and trifunctional phenol [36]. During 1940s, Flory introduced the terms "degree of branching" and "highly branched species" to discuss the molecular weight distribution for three-dimensional polymers in the state of gelation. Later in 1952, Flory theoretically established that the synthesis of highly branched polymers is possible without the risk of any gelation by polycondensation of monomers having AB_x functional groups ($x \geq 2$), where B is capable of reacting with A [37]. This discovery eventually laid the foundation for hyperbranched polymers nearly 30 years back. In 1982, Kricheldorf and his coworkers prepared highly branched copolymers by reacting AB- and AB_2-type monomers in a single step [38]. Finally, with the discovery made by Kim and Webster, the investigation on hb polymers took a flight, and through several courses of trials and modifications finally became technically feasible (Fig. 1.6).

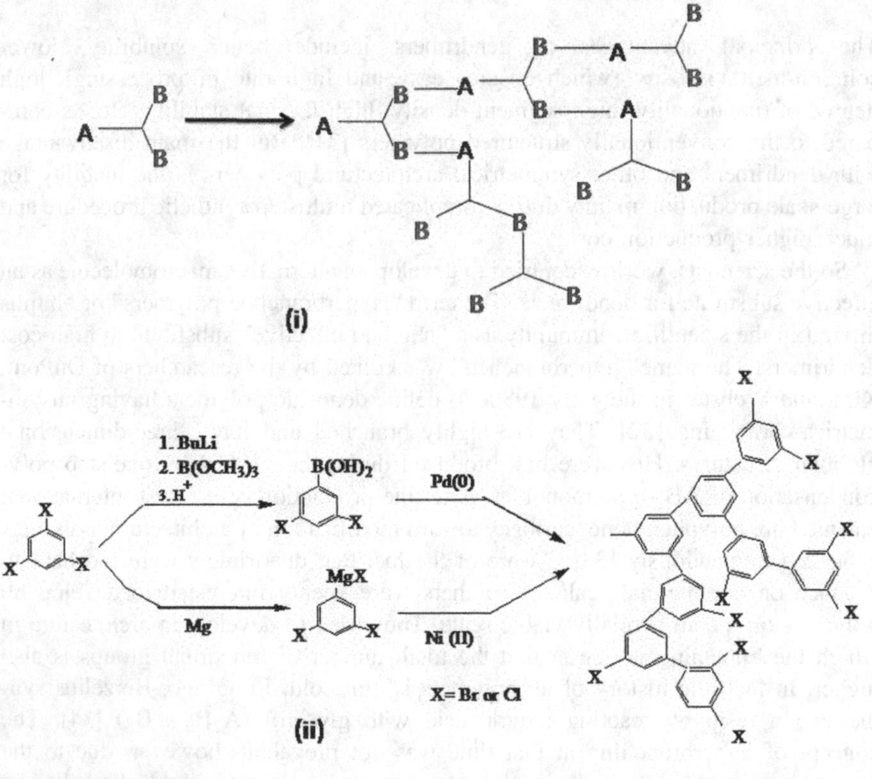

Fig. 1.6 **i** Theoretical model of synthesis of highly branched polymers from polycondensation of AB_2 monomer according to Flory, **ii** Hyperbranched polyphenylene synthesized by Suzuki polycondensation of AB_2 monomer as designed by Kim Webster

Interesting features like highly branched topological structures, the presence of a large number of voids, numerous terminal functional groups, and convenient synthetic procedures of hb polymers render a staggering analogy with dendrimers. Due to their unique and alterable physical/chemical properties, applications of hb polymers have been explored in a large variety of fields [39]. However, some differences still exist specifically in the structural aspects which are pointed out in Table 1.1.

Even though both dendrimers and hb polymers are highly branched, the latter is less symmetrical in molecular structure. The DB of dendrimer is equal to 1, while hb polymers show much lower DB. Hb polymers involve a simple one-pot reaction and avoid the complicated synthesis and purification procedures, which make the large-scale manufacturing process more convenient, thus reducing the production cost compared to the dendrimer.

Invasion of polymers as biomaterials thrived the field of application biotechnology for the last few decades. Polymers with high degradability and biocompatibility are of special interest to biotechnologists because it had widened the application spectrum. Such polymers were able to be broken down and removed after they served their function. Applications are wide ranging, starting from surgical sutures, implants, and manufacture of different medical equipments to the latest drug administration and therapy. Such kind of special polymers must satisfy some important physical, chemical, biological, biomechanical, and degradation properties, specific to application. In a more explicit term, such polymers (1) must encourage cell attachments and growth; (2) should not evoke a sustained inflammatory response; (3) must possess a degradation time coinciding with their function; (4) must produce nontoxic degradation products that can be readily reabsorbed or excreted; (5) must include appropriate permeability and processability for

Table 1.1 Points of contrasts between dendrimers and hb polymer

Dendrimers	Hyperbranched polymers
Dendrimers are uniformly branched, three-dimensional tree-like structured globular-shaped macromolecules	HBPs are globular highly branched macromolecules but exhibit irregularity in structure and branching
Dendrimers have degree of branching equal to 1	HBPs show degree of branching less than 1
Dendrimers can be synthesized via multistep process	HBPs can be synthesized via one-pot synthesis technique
The synthesis steps for dendrimers are tedious and costly	The synthesis processes for HBPs are less complicated and quite cost-effective
The yield for dendrimers is quite low so it is not feasible to produce in large scale	Final yield is quite high enough to be produced in large scale

designed application; (6) should have void structure to be applied for various host–guest applications; (7) should achieve appropriate mechanical properties for specific end use; (8) must be capable of attachment with other molecules (to increase scaffold interaction with normal tissue), and (9) finally should be cost-effective and easy to design [40]. The requirements are huge and some cases are conflicting and thus often not achievable in a single polymer or a single architecture. Or even a single polymer family just cannot satisfy all desirables and thus no such biopolymer family exists either. Instead, a library of polymers has been sorted out by the researchers that either intrinsically or extrinsically engineered into a satisfactory biopolymer for specific applications both in vitro and in vivo stages. The three most intrinsic structural characteristics of an hb polymer are as follows: (i) the predominant disentangled three-dimensional architecture, (ii) greater solubility yet low viscosity, and (iii) large numbers of functional end groups, which lately has attracted the biotechnologists to select such architectural polymers including dendrimers for various biological applications. This particular interest has evoked the idea of developing more and more intrinsic hb polymers meeting specific biological requirements. In addition to that, hb polymers have been synthesized with high biocompatibility and biodegradability, higher ability to control the biological responses, and much easier to incorporate guest molecules within its void spaces through covalent or noncovalent bonding; which of late has made them almost indispensable for many sophisticated biological applications. To date, a significant progress has already been made for the hb polymers in solving some of the very fundamental and technical problems in biological systems [41].

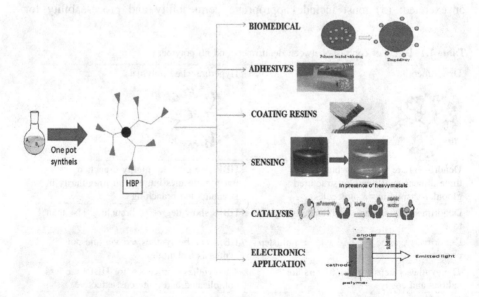

Fig. 1.7 Different applications of hyperbranched polymer

However, in the last two decades, hb polymers have also gained equal importance in material applications such as in surface coating, adhesive, additive in
polymer compounds, and as sensors, catalysts, construction chemical, holography,
cosmetics, and optoelectronics particularly because of its highly flexible branches
and more number of end group functionalities [42]. Following schematic (Fig. 1.7)
summed up these application domains of hb polymers in a single layout.

1.5   Conclusion

Hb polymers have become the hottest topic of research in the field of polymer
science and technology in the last few years as observed from the increasing
number of publications. Due to its uniqueness in properties, high reactivity of the
terminal groups, and wide range of potential applications, hb polymers have drawn
the attention of a major class of scientists and technologists around the world. Till
now, different families of hb polymers have been developed such as polyester,
polyamide, polyether, polyacrylate, polystyrene, and polyolefin, for various uses.
We would like to discuss the synthesis strategies of some of those in the subsequent
chapters with special reference to specific, high-end, and latest bioapplications.

References

1. Young R (1986) Wiley, New York
2. Greve H (200) Ullmann's encyclopedia of industrial chemistry, Wiley
3. Mann CC (2013) 1493: uncovering the new world columbus created, 244–245
4. American Chemical Society National Historic Chemical Landmarks, Bakelite: The World's
   First synthetic plastic (1993)
5. American chemical society national historic chemical landmarks, foundations of polymer
   science: Wallace Carothers and the Development of Nylon, (2000)
6. American Chemical Society National Historic Chemical Landmarks, U.S. Synthetic Rubber
   Program, (1998)
7. American chemical society international historic chemical landmarks, foundations of polymer
   science: Hermann Staudinger and Macromolecules, (1999)
8. Carothers W (1936) Polymers and polyfunctionality. Transaction of the Faraday Society,
   39–49
9. Rubinstein M, Colby RH (2013) Polymer physics. Oxford University Press, Oxford, New
   York, 6
10. Tomalia DA, Naylor AM, Goddard WA (1990) Angew Chem Int Ed Engl 29:138–175
11. Bovey FA, Schilling FC, McCrackin FL, Wagner HL (1976) Macromolecules 9:76–80
12. Yan D, Wang WJ, Zhu S (1999) Polymer 40:1737–1744
13. Compare Materials: HDPE and LDPE. Makeitfrom.com
14. Painter PC, Coleman MM (1997) Fundamentals of polymer science, CRC Press, 14
15. Hadjichristidis N, Pitsikalis M, Iatrou H, Driva P, Sakellariou G, Chatzichristidi M (2012)
    Polymer science: a comprehensive reference, 29–111
16. Milner ST (1991) Science 251:905–914
17. Gary WE (2014) Branched polymer

18. Ruan JZ, Litt MH (1986) Synth Met 15:237–242
19. Hawker CJ, Lee R, Frechet JMJ (1991) J Am Chem Soc 113:4583
20. Shroff RN, Mavridis H (1999) Macromolecules 32(25):8454–8464
21. Tomalia DA, Baker H, Dewald J, Hall M, Kallos G, Martin S, Roeck J, Ryder J, Smith P (1985) Polym J 17:117–132
22. Hodge P (1993) Nature 362:18–19
23. Hawker CJ, Fréchet JMJ (1990) J Am Chem Soc 112:7638–7647
24. Grayson SM, Frechet JM (2001) Chem Rev 101:3819–3868
25. Frechet JMJ (1994) Science-AAAS-Weekly Paper Edition-including Guide to Scientific Information 263(5154):1710–1714
26. Mingjun L, Fréchet JMJ (1999 Pharmaceutical science & technology today, 393–401
27. Deyue Y, Gao C, Frey H (2011) Hyperbranched polymers: synthesis, properties, and applications: John Wiley & Sons, 8
28. Gauthier M (2007) J Polym Sci Part A: Polym Chem 45:3803–3810
29. Teertstra SJ, Gauthier M (2004) Prog Polym Sci 29:277–327
30. Feng Z, Chen Y, Lin H, Wang H, Zhao B (2008) Carbohyd Polym 74:250–256
31. Graham S, Cormack PAG, Sherrington DC (2004) Macromolecules 38:86–90
32. Kim YH, Webster OW (1992) Macromolecules 25:5561
33. Kim YH, Webster OW (1990) J Am Chem Soc 112:4592–4593
34. Zhang X (2010) Prog Org Coat 69:295–309
35. Kienle RH, Hovey AG (1929) J Am Chem Soc 51:509
36. Odian G (1991) Principles of Polymerization 2:96–97
37. Flory PJ (1952) J Am Chem Soc 74:2718
38. Kricheldorf HR, Zang Q, Schwarz G (1982) Polymer 23:1821–1829
39. Voit BI, Lederer A (2009) Chem Rev 109:5924–5973
40. Majumdar J, Edmond L (2006) Australasian Biotechnology 16:14
41. Zhou Y, Huang W, Liu J, Zhu X, Yan D (2010) Adv Mater 22:4567–4590
42. Yates CR, Hayes W (2004) Eur Polymer J 40:1257–1281

Chapter 2
Part I—Synthesis of Hyperbranched Polymers: Step-Growth Methods

List of Abbreviations

DB	Degree of branching
DP	Degree of polymerization
FRP	Free radical polymerization
Hb	Hyperbranched
M.W	Molecular weight
M.W.D	Molecular Weight Distribution
NMP	N-methyl-2-pyrrolidinone solvent
P.D.I	Poly dispersity index
PC	Polycarbonate
PG	Poly glycerol
ROP	Ring opening polymerization

2.1 Introduction to Theoretical Approaches in Hyperbranched Polymerization

Following the extensive works on dendrimers which are structurally perfect but tedious to prepare, the need for the development of structurally imperfect hyperbranched (hb) polymers has gained momentum. A dendrimer is constituted of terminal units (at the globular surface) and dendritic units (inside the macromolecular framework). Whereas a hb polymer is constituted of terminal units (at the irregular surface), linear units and dendritic units (both of which are distributed randomly inside the macromolecular framework). These structural variations in dendrimers and hb polymers arise from the difference in synthesis strategies and mechanism of their formation. A lot of research has already been done and also ongoing to introduce new synthesis strategies for the development of hb polymers for commercialization. Hb polymers may be prepared via one of the three one pot, low-cost pathways- (1) bottom up approach (polymerization of monomers), (2) top down approach (breakdown of macromolecules) and (3) from polymer precursor

© Springer Nature Singapore Pte Ltd. 2018
A. Bandyopadhyay et al., *Hyperbranched Polymers for Biomedical Applications*,
Springer Series on Polymer and Composite Materials,
https://doi.org/10.1007/978-981-10-6514-9_2

molecules [1]. Among these, the bottom up approach is popular in the synthesis of hb polymers which is further categorized into four subdivisions-1) AB_x polycondensation (where $x \geq 2$), (2) vinyl polymerization, (3) $A_2 + B_3$ polymerization; following Flory's rule of equal reactivity and (4) polymerization of asymmetric monomer pairs; following the rule of non-equal reactivity. It is well established from our Nature that branching is an important phenomenon as it facilitates fast and efficient transfer and distribution of energy and mass. Hence, hb polymers would undoubtedly attract biomedical applications where transport phenomenon is one of the essential parts. In fact, there is a constant hunt for the new monomers to develop hb polymers with better tunable properties (say physiochemical properties, biodegradability, biocompatibility, self-assembling properties, peripheral functionality for target specific delivery applications, stimuli responsiveness, etc.) and controlled topologies than the existing ones, especially for the biomedical applications. This chapter is mainly focused on the synthesis of bio-medically important hb polymers (either established or with potential future prospects) through AB_x polymerization and $A_2 + B_3$ polymerization which may follow either step-growth or chain-growth routes.

2.2   Hyperbranched Polymers from AB_x-Type Monomers

Both in theory and practice, majority of the hb polymers with a multitude of functional end groups have been synthesized via one-pot step-growth reaction of AB_x-type monomers (where $x \geq 2$), following either single monomer methodology; SMM or double monomer methodology; DMM. Till today, AB_2, AB_3, AB_4, and AB_6 type monomers have successfully yielded hb polymers [2]. Polycondensation of AB_x-type monomers produces random hb polymers unlike ideal generations of dendrimers, as in the former case the polymer chain formation occurs via random reaction sequences (via dimers, trimers, short/ long oligomers, etc.). The simplest and the most successful hb polymers are generally obtained from AB_2-type monomers (trifunctional monomers), following Flory's cascade theory (whereby crosslinking can be prevented as detailed below) [3]. In an AB_2 polycondensation reaction, if both the B groups of one molecule react with the A group of two different molecules then only a branch unit (a three-arm structure) is generated (see Scheme 2.1). Otherwise, in the same case, if only one of the B groups of one molecule reacts with the A group of another molecule then a linear polycondensed polymer is generated. In an ideal hb polycondensed AB_2 polymer, each molecule definitely contains at most, one unreacted A group (provided there is no intramolecular condensation reaction between A and B) and n + 1 unreacted B groups for n mer units.

However, often polycondensed AB_2 polymers get gelled and hb polymers cannot be obtained owing to the uncontrolled growth in three dimensions. Gelation in an AB_2 system occurs when a critical number of intermolecular linkages get exceeded which must be avoided in order to generate hb polymers [4]. Successful

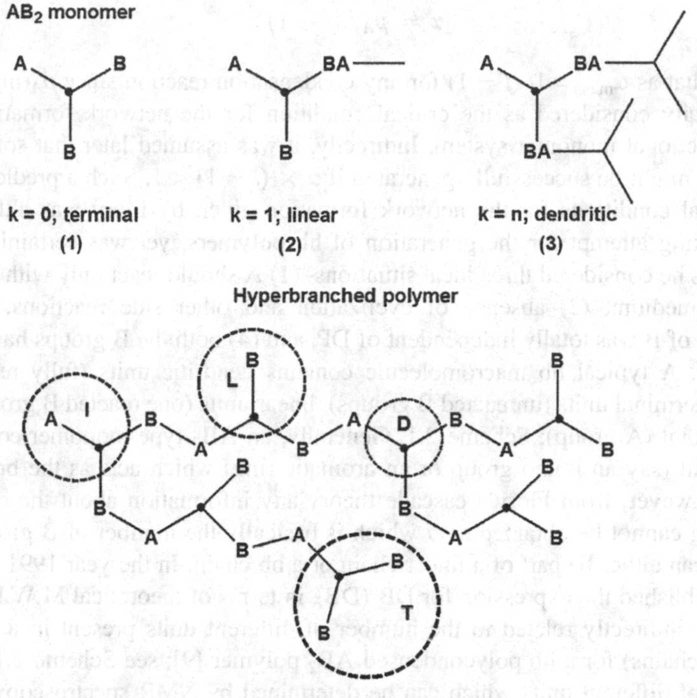

AB$_2$ monomer

k = 0; terminal   k = 1; linear   k = n; dendritic
(1)   (2)   (3)

Hyperbranched polymer

Scheme 2.1 Schematic representation of the probable architecture of a hb polycondensed AB$_2$ polymer; I- initiator unit, L- linear unit, D- dendritic unit, and T- terminal unit

branching theory for the conventional hb polycondensed AB$_2$ polymers was theoretically established by many groups among which those determined by Flory [5], Holter–Frey [6], Moller et al. [7], and Hult et al. [8] are well accepted. Traditional Flory's cascade/ branching theory (in terms of the critical extent of reaction; conversion of B groups) predicted the critical branching coefficient (i.e., the probability for the attachment of a B group of a branch unit to another unit) for a polyfunctional condensation system employing A-R-B$_{f-1}$ monomer where f = number of functional groups per monomer of the same reactivity; $f \geq 3$. Flory assumed,

$$\alpha = p_B \tag{1}$$

where p_B was the fraction of B groups that condensed (provided all the B groups were equally reactive). As A and B react stoichiometrically,

$$p_B \times (f - 1) = p_A \tag{2}$$

where p_A was the fraction of A groups that reacted.

Thus, replacing p_B, Flory obtained,

$$\alpha = p_A / (f - 1) \tag{3}$$

Now that as $\alpha_{max} = 1/(f-1)$ for any condensation reaction since $p_A(max) = 1$, was actually considered as the critical condition for the network formation in a multi functional monomer system. Indirectly, it was assumed later that soluble hb polymers might be successfully generated if $\alpha \times (f-1) < 1$. Such a prediction for the critical conditions for the network formation given by Flory was although a pathbreaking attempt for the generation of hb polymers, yet was certainly hypo- thetical as he considered three ideal situations- (1) A should react only with B in the reaction medium, (2) absence of cyclization and other side reactions, (3) the reactivity of B was totally independent of DP, and (4) both the B groups have equal reactivity. A typical hb macromolecule contains dendritic units (fully reacted B groups), terminal units (unreacted B groups), linear units (one reacted B group), and a focal point (A group); Scheme 2.1. Generally, an AB_2-type monomer contains a focal point (say an imino group or an aromatic ring) which acts as the branching point. However, from Flory's cascade theory any information about the extent of branching cannot be obtained as α which is basically the number of B groups that reacted, can either be part of a linear chain or a hb chain. In the year 1991, Frechet et al. established the expression for DB (DB_F in terms of theoretical M.W.D which in turn is indirectly related to the number of different units present in a mass of polymer chains) for a hb polycondensed AB_2 polymer [9]; see Scheme 2.1 for the location of different units which can be determined by NMR spectroscopy.

$$DB_F = (\Sigma D + \Sigma T)/(\Sigma D + \Sigma T + \Sigma L) \tag{4}$$

Again, in the year 1997, Frey and Holter provided another expression for DB (DB_{HF} in terms of DP) for the AB_2 system [6];

$$
\begin{aligned}
DB_{HF} &= \Sigma D / \Sigma (D)_{max} = 2 \times \Sigma D/(2 \times \Sigma D + \Sigma L) = (\Sigma D + \Sigma T - \text{number of molecules})/ \\
&\quad (\Sigma D + \Sigma T + \Sigma L - \text{number of molecules}) \\
&= 2 \times (\text{avg no. of branching points per hb molecule})/ \\
&\quad (\text{avg DP} - 1)
\end{aligned} \tag{5}
$$

The group predicted that an AB_2-type monomer with similar reactive both B groups yield hb polymers having maximum DB of 0.5 due to the statistical nature of an AB_2 polymerization reaction. Finally, Moller's group further modified Frechet and Frey–Holter branching parameters and established a corelation between DB and the conversion of A groups in an AB_2 system [7]. Moller observed that the value of DB_F decreased from 1.0 (for $p_A = 0$) to 0.5 (for $p_A = 1$) but never decreased below 0.5. Moller also observed that DB_{HF} increased from 0 (for $p_A = 0$) to 0.5 (for $p_A = 1$). In this regard, Frey–Holter parameter provided a better understanding of the extent of branching in a hb AB_2 polycondensate than Flory and Frechet parameter as DB_{HF} increased with the conversion of A group which indirectly suggested the addition of more and more molecules to the branched structures at the branching points. However, Frechet parameter again suffers from the assumption of a generation of monodisperse hb polycondensed AB_2 polymer (an ideal situation). In the present day, DB is considered as one of the important parameters for the determination of the structure of a hb polymer. There are numerous analytical techniques from which DB for an AB_x polycondensate system may be determined (either directly or indirectly) [10] which are detailed in the subsequent Chap. 4 under the Sect. 4.2.

To generate AB_2 hb structures, AB_2 or latent AB_2-type reactants may be used solely as the monomer, AB_2-type monomers may be used as the additive seed in co-polycondensation reactions or AB_2 polycondensation may be carried out in the presence of a third molecule core B_f, where $f \geq 2$ [11]. The first hb polycondensed AB_2 polymer was commercialized under the trade name "Boltorn" by Berzelius (Perstorp Polyols Inc. USA) from the esterification reaction of 2,2 dimethylolpropionic acid. Following Berzelius' work, numerous hb polycondensed AB_2 polymers (from different classes of condensation polymers) have been prepared through controlled condensation reactions, of which some major works include polyarylenes-Suzuki coupling, polyaryleneacetylenes-Heck reaction, polyaryleneether/polyetherketones-aromatic substitution, polycarbosilanes/polycarbosiloxanes-hydrosilylation, etc. Even hb grades of polyesters, polyamides, polyethers, polyethersulfones, polycarbonates, polysiloxanes, polyphenyleneoxides, polyphenylenesulfides, and poly (bis-(alkylene) pyridinium)s.

2.2.1   Carbon–Carbon Coupling Reactions

2.2.1.1   Transition Metal Mediated C–C Coupling Reactions

One of the oldest techniques to synthesize new architectural macromolecules like hb polymers was relied on C–C bond formation, catalyzed by transition metals (mostly Pd and Ni). The first substitute to dendrimers, hb polyphenylenes, **6** was in fact developed by Kim and Webster through a one-step coupling reaction (a Suzuki type coupling reaction) between 3,5 dibromophenyl boronic acid (an unstable AB_2-type monomer intermediate, **4**), in the presence of catalytic amount of tetrakis (triphenylphosphine) Pd (II) in aqueous carbonate (Scheme 2.2) [12]. Alternately, successful hb polyphenylenes, **6** were also prepared from mono-Grignard compound (another unstable AB_2-type monomer intermediate, **5**), in the presence of catalytic amount of tetrakis(phenylphosphine) nickel chloride.

Polyphenylenes are the typical examples of transition metal catalyzed chemo-selective aryl–aryl coupling reactions. Unlike linear polyphenylenes, hb polyphenylenes are soluble in many organic solvents like THF, o-dichloro benzene, tetrachloroethane, etc., and thus processable owing to the presence of numerous voids spaces (where solvent molecules may be entrapped). Void spaces are generated from irregular microstructures of the hb polymers and also for reduction in

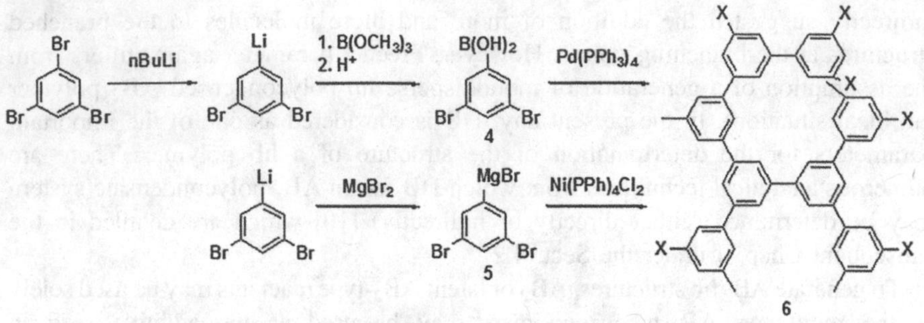

4 & 5 are AB$_2$ intermediates; 6 is the hyperbranched polyphenylene

Scheme 2.2 Scheme showing transition metal catalyzed C–C coupling reaction for the generation of hb polyphenylenes [12]

$\pi - \pi$ stacking interactions [13]. Hb polyphenylenes are basically nonconducting polymers because extended π conjugation is prevented in the microstructure due to the presence of packed and twisted phenylene units and hence they may be suitable as high-performance insulators. Diverse derivatives of functionalized hb polyphenylenes were also prepared by many researchers by electrophilic substitution reactions at positions bearing halogen groups either on polylithio-phenylenes or polyphenylenes, with electrophilies like $-CO_2$, CH_3OCH_2Br, $(CH_3)_2O$, ROH, DMF, etc. Often with the introduction of hydrophilic electrophiles, hb polyphenylenes become water soluble and get suitable in many bio-based applications owing to the rendered biocompatibility and biodegradability. In an attempt to explore the world of aromatic hb polymers, Tanaka et al. synthesized hb poly (triphenylamine)s, **8** by Ni (II) catalyzed coupling of an AB$_2$-type Grignard reagent, **7** (Scheme 2.3) [14]. However, the intermediate AB$_2$-type monomer, **7** is highly

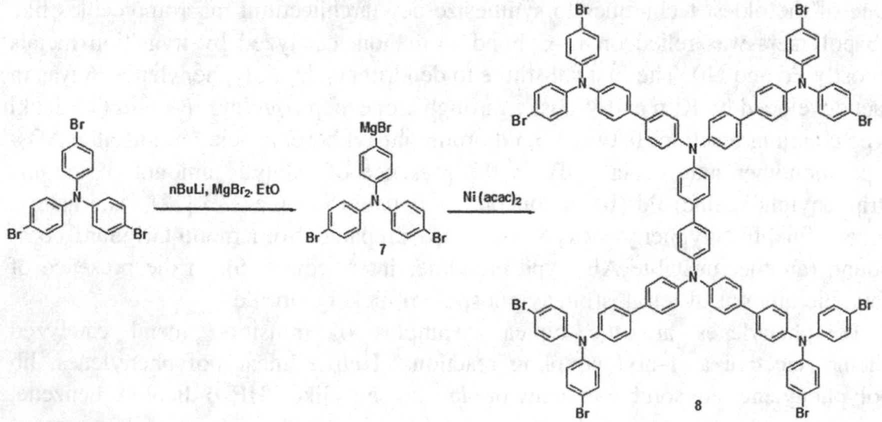

Scheme 2.3 Scheme showing Ni (II) catalyzed C–C coupling reaction for the generation of hb poly (triphenylamine)s [14]

moisture/air sensitive and hence restricted the production of hb polymer, **8** for commercial purposes.

So far, most of the successful formation of hb polymers through transition metal catalyzed C–C coupling reaction was carried out either by Suzuki coupling reaction or by Heck coupling reaction. Suzuki coupling reaction was employed in the synthesis of hb polyarylenes, poly (triphenylamine)s and poly (aryleneether)s. Sun et al. prepared hb poly (triphenylamine)s, **10** from 4-[bis(4-bromophenyl)amino] benzene boronic acid in its ester form, **9**; a derivative of tris(4-bromophenyl)amine, by a Suzuki coupling polycondensation reaction in a one-pot system; Scheme 2.4 [15]. The same group also synthesized an alternating hb copolymer of poly (triphenylamine-fluorene), **12** of higher M.W and better solubility than the hb poly (triphenylamine)s-homo polymer, by a Suzuki coupling polycondensation reaction between **9** and a multifunctional core molecule named as 2,7-diiodo-9,9-dioctylfluorene, **11**; Scheme 2.4 [15].

Hb aryl/alkyl polymers with controlled architecture and high DB were prepared by Bo et al. from an AB_2-type monomer, in the presence of an AC_2-type monomer (where AC_2 is more reactive than AB_2 under mild temperature conditions) by a one-pot Suzuki coupling polycondensation reaction; Scheme 2.5 [16]. A two-step temperature variation was followed to ensure a controlled growth of the dendritic

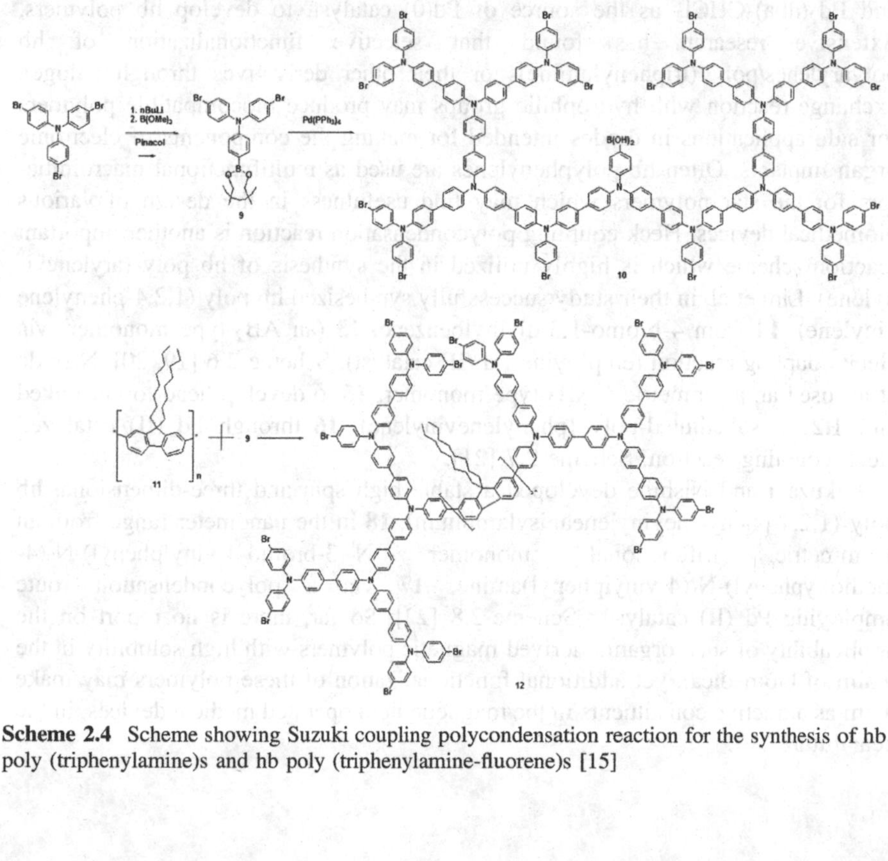

Scheme 2.4 Scheme showing Suzuki coupling polycondensation reaction for the synthesis of hb poly (triphenylamine)s and hb poly (triphenylamine-fluorene)s [15]

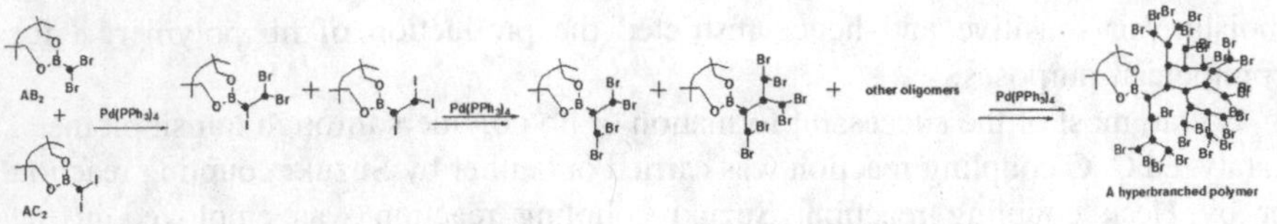

Scheme 2.5 Schematic representation of a probable mechanism for the formation of a hb polymer through AB_2 + AC_2 type Suzuki coupling polycondensation reaction; there was a difference in the reactivity of the monomers under different reaction conditions [16]

structures. AC_2; an iodo-aromatic compound reacted faster than AB_2; a bromo-aromatic compound with boronic esters, at a lower temperature and hence formed multi B functional AB_n-type branched oligomers. In the further step, under refluxing condition, AB_n-type oligomers polymerized to form highly branched structures. Often catalyst transfer reactions are favored to improve DB of the hb polymers. Huang et al. synthesized a hb polymer with a very high DB-100% [17]. There are reports on the improvement of DB of the hb polymers but they are less explored as there are only a few monomers which may allow such achievement [18]. The group of Huang used an AB_2-type monomer containing one aromatic boronic pinacol ester and two aromatic bromo groups linked by an alkyl chain spacer for the catalyst transfer polymerization in the presence of $P(t-Bu)_3$; the ligand and $Pd_2(dba)_3CHCl_3$ as the source of Pd(0) catalyst, to develop hb polymers. Extensive research has found that selective functionalization of hb polyarylenes/poly (triphenylamine)s or their other derivatives through halogen exchange reaction with hydrophilic groups may produce biocompatible polymers for safe applications in diodes intended for making the components of electronic organ implants. Often hb polyphenylenes are used as multifunctional macroinitiators for the star polymers which may find usefulness in the design of various biomedical devices. Heck coupling polycondensation reaction is another important reaction scheme which is highly utilized in the synthesis of hb poly (arylenevinylene). Lim et al. in their study successfully synthesized hb poly (1,2,4-phenylene vinylene), **14** from 4-bromo-1,3-divinylbenzene, **13** (an AB_2-type monomer) via Heck coupling reaction (employing Pd (II) catalyst); Scheme 2.6 [19, 20]. Nishide et al. used an asymmetrical AB_2-type monomer, **15** to develop head to tail linked and 1,2,4,6 substituted-poly (phenylenevinylene), **16** through Pd (II) catalyzed Heck coupling reaction; Scheme 2.7 [21].

Fukuzaki and Nishide developed a stable high spin and three-dimensional hb poly (1,2,4-phenylenevinyleneanisylaminium), **18** in the nanometer range from an asymmetric trifunctional monomer N-(3-bromo-4-vinylphenyl)-N-(4-methoxyphenyl)-N-(4-vinylphenyl)amine, **17** via a polycondensation route employing Pd (II) catalysts; Scheme 2.8 [22]. So far, there is no report on the applicability of such organic-derived magnetic polymers with high solubility in the realm of biomedical, yet additional functionalization of these polymers may make them as attractive constituents in the magnetic field operated medical devices, in the near future.

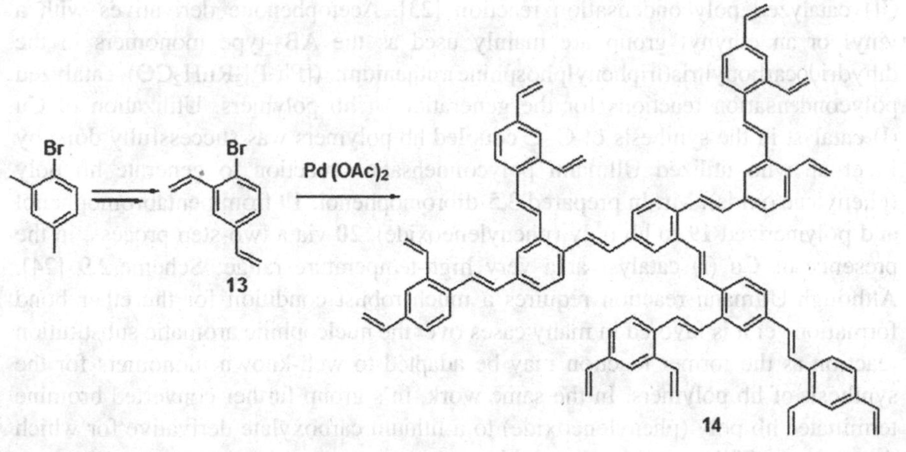

Scheme 2.6 Scheme showing Heck coupling reaction for the synthesis of hb poly (1,2,4-phenylenevinylene) [19]

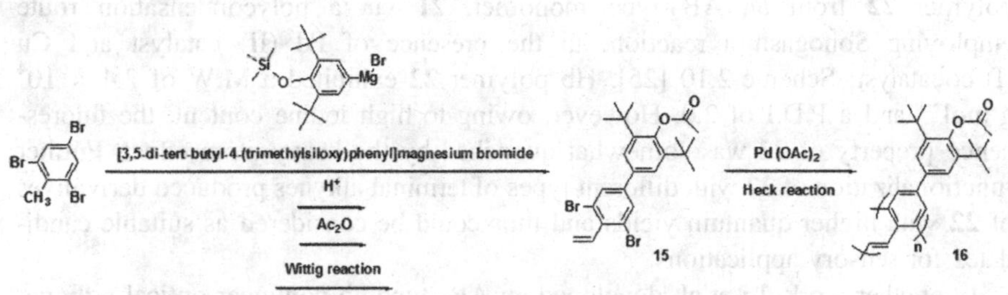

Scheme 2.7 Scheme showing Heck coupling polycondensation reaction for the synthesis of hb 1,2,4,6 substituted- poly (phenylenevinylene) [21]

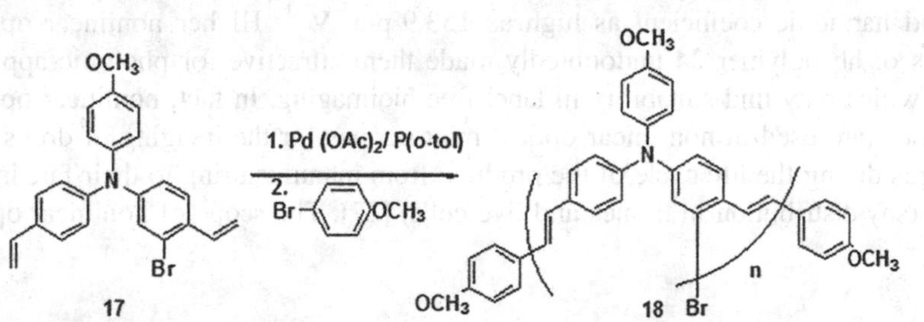

Scheme 2.8 Scheme showing Heck coupling polycondensation reaction for the synthesis of hb poly (1,2,4-phenylenevinyleneanisylaminium) [22]

Other transition metals that are often used as the catalysts in the synthesis of AB₂ hb polycondensed polymers include ruthenium and copper. Lu et al. reported successful generation of hb poly (4-acetylstyrene); an AB₂-type monomer via Ru

(II) catalyzed polycondensation reaction [23]. Acetophenone derivatives with a vinyl or an ethynyl group are mainly used as the AB_2-type monomers in the dihydridocarbonyltris(triphenylphosphine)ruthenium; ([Ph₃P]₃RuH₂CO) catalyzed polycondensation reactions for the generation of hb polymers. Utilization of Cu (I) catalyst in the synthesis of C–C coupled hb polymers was successfully done by In et al. who utilized Ullmann polycondensation reaction to generate hb poly (phenylene oxide). Mr. In prepared 3,5-dibromophenol, **19** from pentabromophenol and polymerized **19** to hb poly (phenyleneoxide), **20** via a two-step process, in the presence of Cu (I) catalyst at a very high-temperature range; Scheme 2.9 [24]. Although Ullmann reaction requires a much robust condition for the ether bond formation yet it is favored in many cases over the nucleophilic aromatic substitution reaction as the former reaction may be adapted to well-known monomers for the synthesis of hb polymers. In the same work, In's group further converted bromine terminated hb poly (phenyleneoxide) to a lithium carboxylate derivative for which the structure **20** became water soluble.

Of the various cross-coupling reactions, recently Sonogashira reactions have gained impetus in the synthesis of hb polymers. Tolosa and his team produced hb polymer **22** from an AB_2-type monomer, **21** via a polycondensation route employing Sonogashira reaction, in the presence of Pd (II) catalyst and Cu (I) cocatalyst; Scheme 2.10 [25]. Hb polymer **22** exhibited a M.W of 2.4×10^4 g mol^{-1} and a P.D.I of 2.0. However, owing to high iodine content, the fluorescence property of **22** was somewhat quenched by the heavy atom effect. Further functionalization of **22** with different types of terminal alkynes produced derivatives of **22** with higher quantum yields and thus could be considered as suitable candidates for sensory applications.

In another work, Li et al. developed an AB_2-type hb nonlinear optical polymer **24** from monomer **23** via Sonogashira coupling reaction; Scheme 2.11 [26]. Hb polymer **24** was soluble in a range of solvents (CHCl₃, THF, DMF, and DMSO), thermally stable, optically transparent and exhibited an M.W of 11,750 g mol^{-1} and second harmonic coefficient as high as 153.9 pm. V^{-1}. Higher nonlinear optical effects of hb polymer **24** undoubtedly made them attractive for photonic applications which may find suitability in label-free bioimaging. In fact, nonlinear optical polymers are used in non linear optical microscopes for the imaging of drugs and dosages during the life cycle of the product, from manufacturing to their fate in the body (say distribution in tissues and live cells) [27]. The scope of nonlinear optics

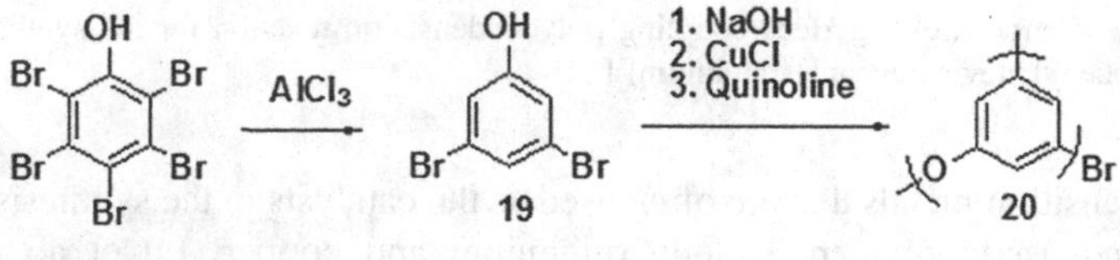

Scheme 2.9 Scheme showing Ullmann polycondensation reaction for the synthesis of hb poly (phenyleneoxide) [24]

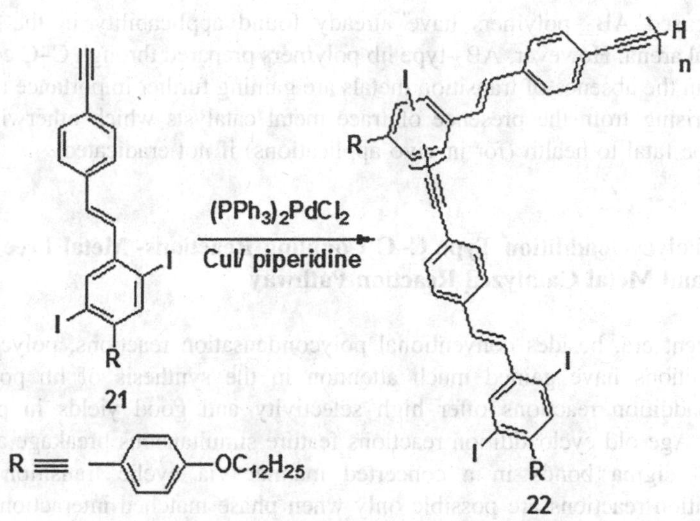

Scheme 2.10 Scheme showing Sonogashira polycondensation reaction for the synthesis of a fluorescent hb polymer [25]

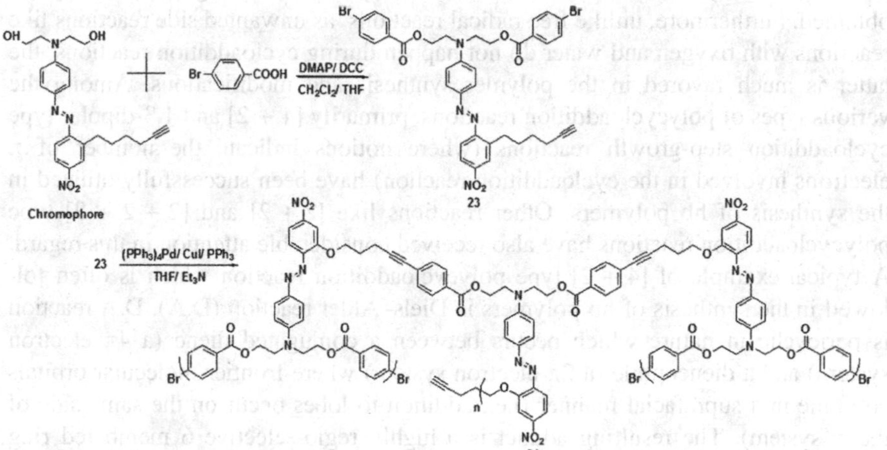

Scheme 2.11 Scheme showing Sonogashira polycondensation reaction for the synthesis of a hb polymer with nonlinear optical properties [26]

has gained much importance especially with the introduction of the concept of personalized therapy.

From the different discussed works, it is observed that transition metal catalyzed C–C coupling polycondensation reactions of AB_2-type monomers leave an indelible effect in the field of hb polymer synthesis. Through the correct choice of monomers and post polymer functionalization with biocompatible moieties, hb polycondensed AB_2 polymers have already found applicability in the diverse biomedical arena. However, AB_2-type hb polymers prepared through C–C coupling reactions in the absence of transition metals are gaining further importance to avoid toxicity arising from the presence of trace metal catalysts which otherwise may prove to be fatal to health (for in vivo applications) if not eradicated.

2.2.1.2   Polycycloaddition Type C–C Coupling Reactions- Metal Free and Metal Catalyzed Reaction Pathway

In the recent era, besides conventional polycondensation reactions, polycycloaddition reactions have gained much attention in the synthesis of hb polymers. Polycycloaddition reactions offer high selectivity and good yields in polymer synthesis. Age old cycloaddition reactions feature simultaneous breakage and formation of sigma bonds in a concerted manner via cyclic transition states. Cycloaddition reactions are possible only when phase matched interactions occur between the highest occupied molecular orbital (HOMO) of the unsaturated end and the lowest unoccupied molecular orbital (LUMO) of the other end, respectively, of the same substrate. Unlike condensation reactions, small molecules are not produced during cycloaddition reactions. Thus, species with high MWs may be obtained. Furthermore, unlike free radical reactions, as unwanted side reactions like reactions with oxygen and water do not happen during cycloaddition reactions, the latter is much favored in the polymer synthesis and modifications. Among the various types of polycycloaddition reactions, primarily [4 + 2] and 1,3-dipolar type cycloaddition step-growth reactions (where notions indicate the number of π electrons involved in the cycloaddition reaction) have been successfully utilized in the synthesis of hb polymers. Other reactions like [2 + 2] and [2 + 2 + 2] type polycycloaddition reactions have also received considerable attention in this regard. A typical example of [4 + 2] type polycycloaddition reaction which is often followed in the synthesis of hb polymers is Diels–Alder reaction (D.A). D.A reaction is pericyclic in nature which occurs between a conjugated diene (a 4π electron system) and a dieneophile (a 2π electron system) where frontier molecular orbitals combine in a suprafacial manner (i.e., addition to lobes occur on the same side of the π system). The resulting adduct is a highly regio-selective 6 membered ring structure. Unlike other polycycloaddition reactions, D.A reaction follows a thermally reversible mechanistic pathway during polymerization. Such thermo-reversibility often makes D.A cycloadducts attractive candidates in thermo-responsive drug/gene delivery devices. However, in practice, D.A reaction often fails to produce polymers with high M.Ws owing to the side reactions and retro [4 + 2] cycloadditions. By employing a strategy of synchronous aromatization and irreversible loss of a small molecule (say carbon monoxide), retro [4 + 2] cycloadditions may be prevented and polymerization is thus favored. After the pioneering work of Stille et al. where they synthesized linear polyphenylenes [28], Morgenroth and Mullen prepared hb polyphenylenes via repetitive inter molecular D.A-based C–C coupling technique, from two kinds of AB_2-type monomers of tetraphenylcyclopentadienones; Scheme 2.12a [29]. In this notable work, they utilized AB$_2$-type monomers comprising of one cyclopentadienone as a diene and two triple bonds as dienophiles. Typically, one equivalent of triisopropylsilyl protected diene, **25** was reacted with two equivalents of tetrabutylammonium fluoride at 180 °C in diphenylether to generate deprotected dienophiles in situ in the existing structure; thereby an AB$_2$-type monomer was developed which finally got polymerized to hb polyphenylenes, **26** in a time period of 12 h. Polymeric cycloadduct **26** was soluble in most of the common organic solvents, exhibited an average M.W of $\sim 17{,}000$ g mol^{-1} and P.D.I of 6.85. The same group also synthesized hb polyphenylenes, **28** from a phenyl substituted in situ generated AB$_2$-type monomer which in turn is obtained from a phenyl substituted diene, **27** in a similar fashion as the previous work; Scheme 2.12b. Polymeric cycloadduct **28** displayed higher average M.W ($\sim 1.07 \times 10^5$ g mol^{-1}) but a lower P.D.I (4.3) than polymer **26**, respectively. Unlike in Pd (II) catalyzed coupling of arylboronic acids with aryl halides where only structures with 1,3,5-linked triphenyl benzene units are produced, D.A reaction gives birth to architectures with densely packed benzene rings. Owing to the dense packing of benzene rings in the hb polyphenylenes prepared by D.A reaction, they undergo intramolecular dehydrogenation reactions which produce poly cyclic aromatic hydrocarbons. As often, poly cyclic aromatic hydrocarbons are used in gene therapy or drug delivery, the method of D.A is preferred in the synthesis of densely packed polyphenylenes over the other methods of preparation for the absence of harmful trace metals.

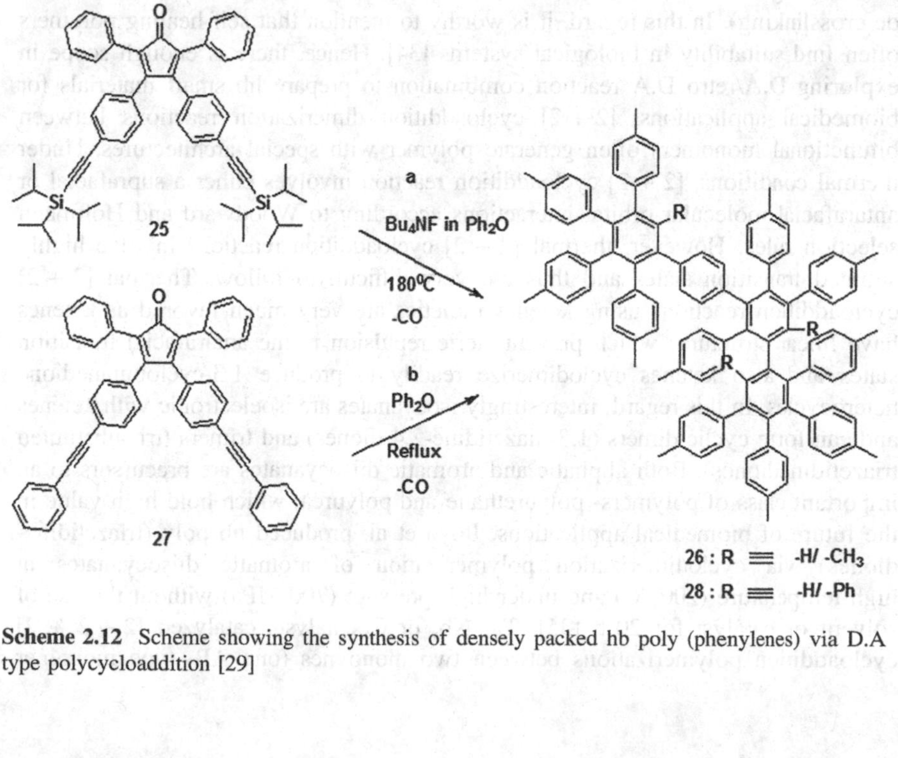

25

27

Bu$_4$NF in Ph$_2$O
180°C
-CO

a

b
Ph$_2$O
Reflux
-CO

26 : R $\equiv$ -H/ -CH$_3$
28 : R $\equiv$ -H/ -Ph

Scheme 2.12 Scheme showing the synthesis of densely packed hb poly (phenylenes) via D.A type polycycloaddition [29]

Following the same methodology, Harrison and Feast prepared a hb polymer constituting of maleimide and cyclopentadienone (i.e., a polyimide) from an AB_2-type monomer via D.A reaction [30]. The resulting hb polyimide was soluble in common organic solvents. Hb polyimides are very useful for encapsulation of biomacromolecules and insulation of active implants [31].

Nowadays, the combination of conventional D.A reaction and retro D.A cycloreversion reaction has gained significant importance in the synthesis of smart materials especially which display thermoresponsiveness with respect to physical properties (like color, viscosity, etc.) [33]. Such reactive polymers are highly useful for biomolecules immobilization, drug/gene delivery, and enzyme modifications. Gok and Sanyal prepared multi arm star polymers containing thiol reactive maleimide groups, **30** at the focal point; Scheme 2.13 [32]. The team deliberately masked double bonds of the reactive maleimide groups with furan via D.A reaction and generated a macroinitiator, which subsequently underwent a living polymerization with various methacrylate and acrylate monomers. Finally, they deprotected maleimide groups at the core of the polymers via retro D.A process (thermal treatment). Such macromolecular multi arm reactive polymeric scaffolds may be conjugated with drug molecules and such systems exhibit enhanced bioavailability and reduced clearance rate. In another work, Froimowicz and his team utilized anthracene functionalized hb polyglycerols in self-healing process via [4 + 4] reversible photo-cyclo-addition reactions [34]. Anthracenes undergo forward [4 + 4] dimerization process (or rather crosslinking in the case of anthracene containing polymers) when irradiated with 366 nm light and finally irradiation with 254 nm light induces backward decomposition of the dimmers (or rather de-crosslinking). In this regard, it is worthy to mention that self-healing polymers often find suitability in biological systems [34]. Hence, there is enough scope in exploring D.A/retro D.A reaction combination to prepare hb smart materials for biomedical applications. [2 + 2] cycloaddition dimerization reactions between bifunctional monomers often generate polymer with special architectures. Under thermal conditions, [2 + 2] cycloaddition reaction involves either a suprafacial or antarafacial molecular orbital interactions according to Woodward and Hoffmann selection rules. However, thermal [2 + 2] cycloaddition reactions involve highly strained transition states and thus are very difficult to follow. Thermal [2 + 2] cycloaddition reactions using ketenes moieties are very much favored as ketenes have linear structures which prevent steric repulsion in the antarafacial transition states and also ketenes cyclodimerize readily to produce 1,3-cyclobutanedione heterocycles. In this regard, interestingly isocyanates are isoelectronic with ketenes and can form cyclic dimers (1,3-diazetidine-2,4-diones) and trimers (tri substituted triazetidinediones). Both aliphatic and aromatic diisocyanates are precursors to an important class of polymers- polyurethane and polyurea; which hold high value in the future of biomedical applications. Itoya et al. produced hb poly (triazetidinediones) via cyclodimerization polymerization of aromatic diisocyanates at high-temperature (200 °C) and under high pressure (700 MPa) without the use of solvent or catalyst for 20 h [35]. Ta, Nb, or Cocatalysts catalyzed [2 + 2 + 2] cycloaddition polymerizations between two monoynes (one AB_2-type monomer

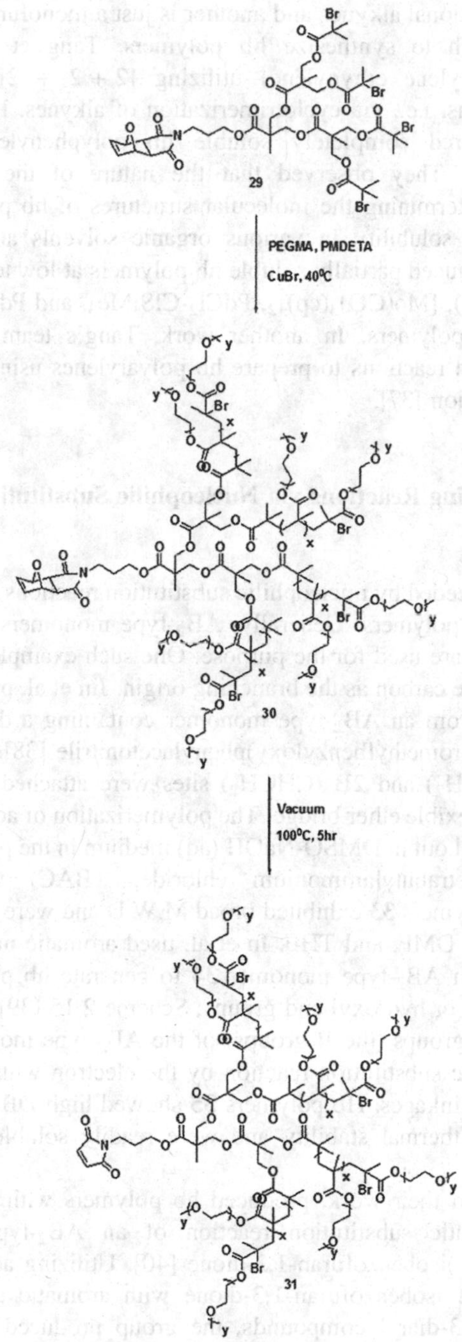

Scheme 2.13 Scheme showing the synthesis of multi arm star polymers containing reactive functional groups via D.A/retro D.A reaction strategy [32]

constituting of bifunctional alkynes and another is just a monofunctional alkyne) is just another approach to synthesize hb polymers. Tang et al. produced hb polyarylenes/poly (arylene ethynylene) utilizing [2 + 2 + 2] inter molecular cycloaddition reactions, i.e., via cyclotrimerization of alkynes. In one such work, Tang's group prepared completely soluble hb polyphenylenes via polycyclotrimerization [36]. They observed that the nature of the catalysts played important roles in determining the molecular structures of hb polymers and thus affected their yields, solubility in various organic solvents and other physical properties. $TaCl_5$ produced partially soluble hb polymers at low temperatures while $NbCl_5$, $Mo(CO)_4(nbd)$, $[Mo(CO)_3(cp)]_2$, $PdCl_2$-$ClSiMe_3$, and Pd/C–$ClSiMe_3$ gave birth to soluble hb polymers. In another work, Tang's team further explored polycyclotrimerization reactions to prepare hb polyarylenes using Cocatalyst activated via UV irradiation [37].

2.2.1.3 C–C Coupling Reactions via Nucleophilic Substitution Reactions

Polymerizations proceeded by nucleophilic substitution reactions are quite useful in the generation of hb polymers. Generally, AB_2-type monomers constituting activated leaving groups are used for the purpose. One such example includes the use of activated methylene carbon as the branching origin. Jin et al. prepared controlled branched polymers from an AB_2-type monomer containing a difunctional nucleophile, **32**; 4-(4'-chloromethylbenzyloxy)phenylacetonitrile [38]. In this monomer **32**, both the A ($ClCH_2$) and 2B ($CHCH_2$) sites were attached separately to the aromatic rings by a flexible ether bridge. The polymerization of activated methylene monomers was carried out in DMSO-NaOH (aq) medium in the presence of a phase transfer catalyst (tetrabutylammonium chloride; TBAC) without gelation; Scheme 2.14. Hb polymers **33** exhibited broad M.W.D and were soluble in organic solvents like DMSO, DMF, and THF. In et al. used aromatic nucleophilic substitution reactions on an AB_2-type monomer **34** to generate hb poly (arylene ether amide)s with fluorine or hydroxyl end groups; Scheme 2.15 [39]. In this work, the two fluorine leaving groups (the B groups) of the AB_2-type monomers were activated for the aromatic substitution reaction by the electron withdrawing carbonyl groups of the amide linkages. Hb polymers **35** showed high DB (0.43–0.53), high Tg (>200 °C), high thermal stability and were readily soluble in aprotic polar solvents.

Yang and Kong in their work, produced hb polymers with a high DB, via a Friedel–Crafts aromatic substitution reaction of an AB_2-type monomer, **36**; 5-(4-phenoxyphenoxy)isobenzofuran-1,3-dione [40]. Utilizing acid-catalyzed condensation reaction of isobenzofuran-1,3-dione with aromatic compounds which exclusively yields 3,3-diaryl compounds, the group produced hb poly (arylene isobenzofuran-1(3H)-one), **37**; Scheme 2.16.

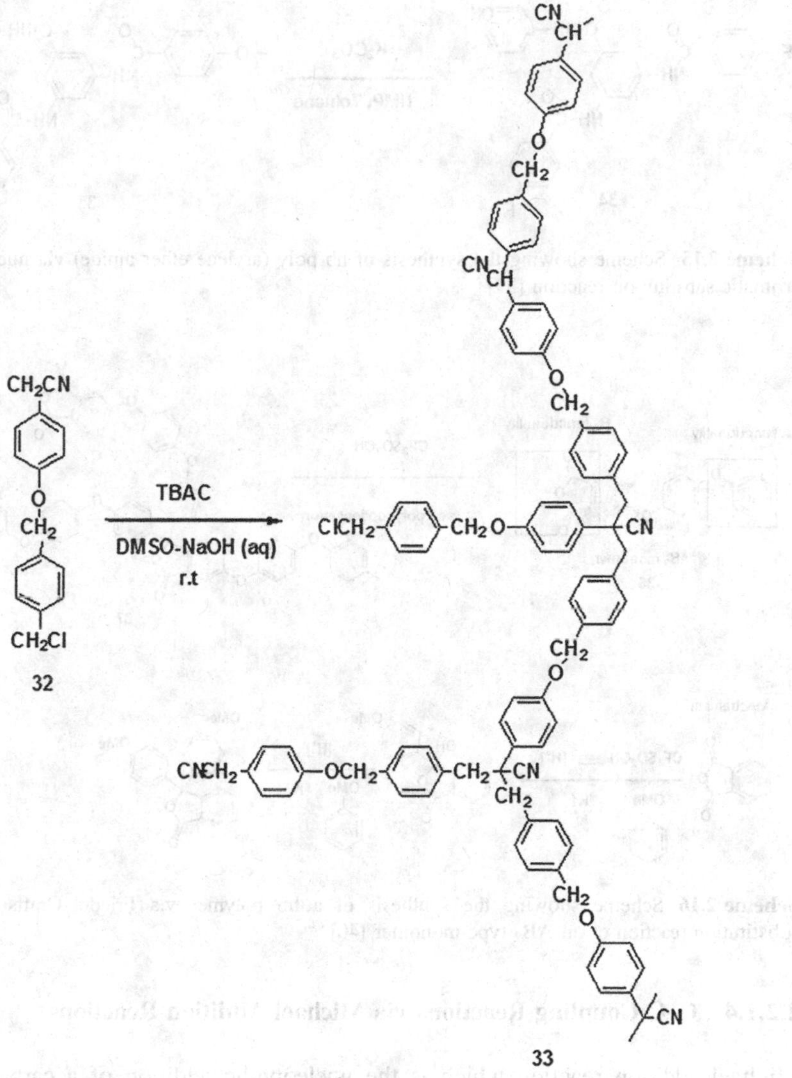

Scheme 2.14 Scheme showing nucleophilic substitution polymerization of activated methylene monomers to hb polymers [38]

So far, we have not obtained any hb polymer synthesized via nucleophilic substitution reactions to be useful in biomedical applications. However, we fervently believe that in future, nucleophilic substitution-based polymerization reactions hold too much prospect in the generation of hb polymers for external medical devices.

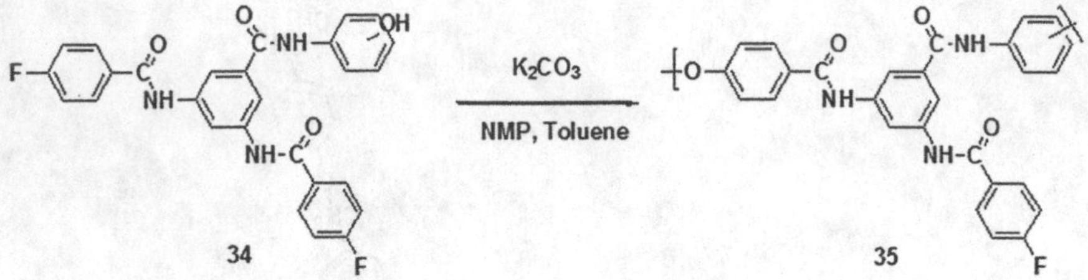

Scheme 2.15 Scheme showing the synthesis of hb poly (arylene ether amide) via nucleophilic aromatic substitution reaction [39]

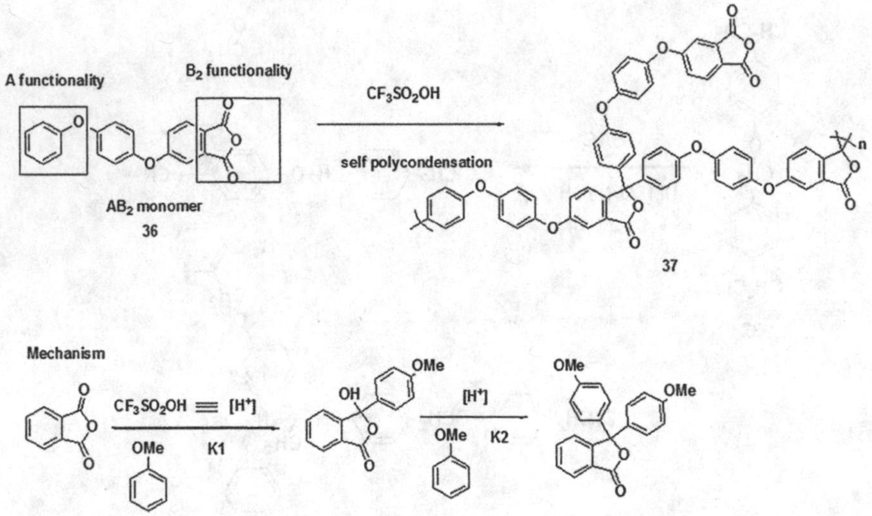

Scheme 2.16 Scheme showing the synthesis of a hb polymer via Friedel–Crafts aromatic substitution reaction of an AB$_2$-type monomer [40]

2.2.1.4   C–C Coupling Reactions via Michael Addition Reactions

Michael addition reaction which is the nucleophilic addition of a carbanion or another nucleophile to an α, ß-unsaturated carbonyl compound, has also attracted the synthesis of hb polymers. Michael addition reaction proceeds rapidly at room temperatures and involves less toxic precursors which often suits biomedical applications. Endo et al. procured an AB$_2$-type monomer, **38**; 2-(acetoacetoxy)ethyl acrylate and carried out Michael addition reaction with **38** in the presence of a mid base 1,8-diazabicyclo-[5.4.0]undec-7-ene (DBU) catalyst, to generate hb poly (ß-ketoester); Scheme 2.17 [41]. Hb polymers **39** exhibited high DB around 0.43–0.829 and were highly soluble in DCM and CHCl$_3$.

D.L Trumbo used Michael addition reaction to polymerize diacrylates (tripropylene glycol diacrylate) and bisacetoacetates to generate hb polymers in the

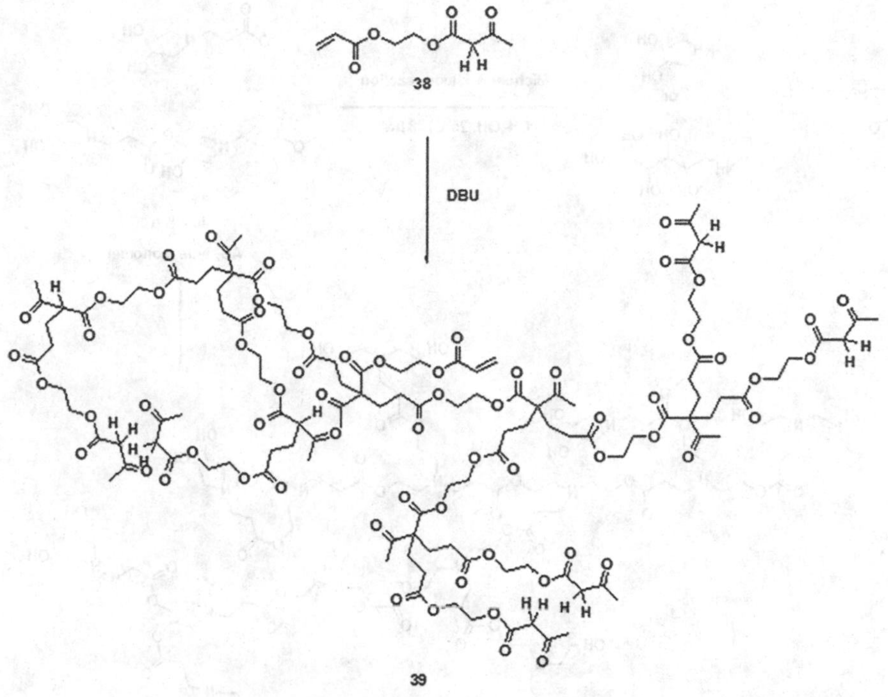

Scheme 2.17 Scheme showing the synthesis of hb poly (ß-ketoester) from an AB$_2$-type monomer via Michael addition reaction [41]

presence of DBU catalysts [42]. Difunctionality of acetoacetate groups facilitated the formation of hb polymers. The hb polymers exhibited high M.W ($\sim 4.37 \times 10^5$ g mol^{-1}) but broad M.W.D (P.D.I $\sim$ 10). Concerning the field of biomedical applications, Gao et al. precisely developed highly water soluble, biodegradable hb polyesters with a large amount of terminal hydroxyl groups from an AB$_2$-type monomer and found them to be suitable for drug delivery; Scheme 2.18 [43]. The group prepared the intermediate AB$_2$/AD$_n$ type monomer (an ester diol, **40**) from diethanolamine and methyl acrylate via Michael addition reaction in situ and continued the polymerization at a higher temperature. The hb polyester, **41** contained tertiary amino groups in the backbone/end hydroxyl groups with moderate M.W and DB was slightly greater than 0.5.

Once again, Park and his team synthesized hb polymer-based gene transfection agents via Michael addition reaction between low M.W linear polyethylenimine, **42** and polyethylene glycol diacrylate, **43**; Scheme 2.19 [44]. The resultant hb polymer **44** was highly branched due to the inherent branching in polyethylenimine and also for the reaction at multiple amine sites along the polymer backbone. The hb polymer **44** was water soluble, biodegradable, exhibited M.W around 10^3 g mol^{-1}, had low cytotoxicity, formed complexes with plasmid DNA, and enabled gene transfection in HepG2/MG63 cell lines with high efficiency.

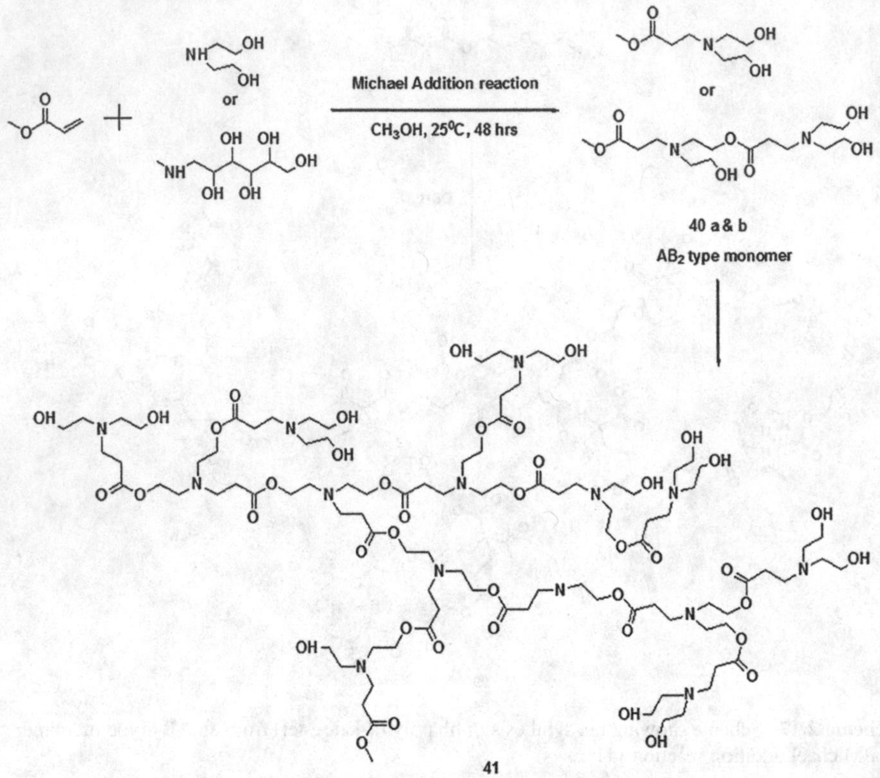

Scheme 2.18 Scheme showing the synthesis of a hb polyester via Michael addition reaction [43]

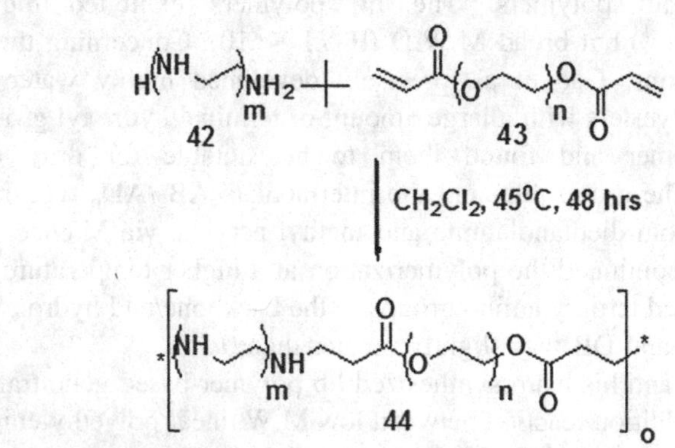

Scheme 2.19 Scheme showing hb copolymer formation between polyethylenimine and polyethylene glycol diacrylate [44]

2.2.2 Carbon-Hetero Atom Coupling Reactions

2.2.2.1 C-N Coupling Reactions via Condensation of Amines and Acid Derivatives

Linear aromatic polyamides and polyimides are known as high-performance polymers owing to excellent thermal, mechanical, and chemical properties. However, these aromatic polymers are insoluble in common organic solvents at room temperature due to the presence of rigid repeating units. Often linear aromatic polymers are restricted in a number of applications because of their robust structures. Introducing dendritic units into the aromatic polymers not only improve solubility but also make them suitable in other sectors of applications including biomedical applications. In fact, hb aromatic polyamides are potentially used as supporting materials for protein immobilization [45]. Hb polyamides, hb polyimides, and hb polybenzoxazoles are generally prepared via condensation of amines and acid derivatives, i.e., via amidation of AB_x-type monomers. It was for the first time Kim introduced liquid crystalline property in the hb aromatic polyamide materials [46]. He prepared hb polyamides (-COOH terminated-**46** and –NH$_2$ terminated-**48**) from AB_2-type monomers, **45** and **47** in an amide solvent at low temperature; Scheme 2.20. Both hb polyamides **46** and **48** exhibited M.W of 2.4–4.6 × 10^3 g mol^{-1}, P.D.I around 2.0–3.2. Hb polyamide **46** displayed nematic phase liquid crystalline properties and did not lose birefringence up to 150 °C. Often nematic liquid crystalline polymers are used in biosensors and ophthalmic lenses [47]. Hence, liquid crystalline hb polymers may be useful as potential biosensors.

Many a times, hb polyamides are synthesized via direct polycondensation reaction in the presence of condensing agents which activate –COOH groups of the AB_x-type monomers in situ. The structure of AB_x-type monomers very much affects DB of the hb polyamides during direct polycondensation reactions. Ishida and his team found that the use of AB_x-type dendritic macromonomers enhances DB of the hb polymers [48]. AB_4, AB_6, and AB_8 type dendrons of amino benzoic acids produced hb polyamides with DB as high as 0.7–0.8 unlike those obtained from AB_2-type monomers (DB $\sim$ 0.32). Direct/self-polycondensation reactions were further explored in the synthesis of hb polyimides in order to ensure high DB. Hb polyimides were prepared by a two-step process, via self-polycondensation of an intermediate polyamic acid methyl ester precursor; Scheme 2.21 [49]. Yamanaka et al. prepared an AB_2-type monomer, **49** which was transformed into polyamic acid methyl ester precursor, **50** in situ using a condensing agent, (2,3-dihydro-2-thioxo-3-benoxazolyl)phosphonic acid diphenyl ester (DBOP) in NMP. Finally, hb poly imide, **51** was synthesized from **50** via chemical imidization process in the presence of acetic anhydride and pyridine in DMSO at 100 °C for 24 h. The resultant hb polyimide was soluble in DMF/DMSO/NMP, had a DB of 0.48, an M.W of 1.8 × 10^5 g mol^{-1} and a P.D.I of 3.0 with outstanding thermal stability.

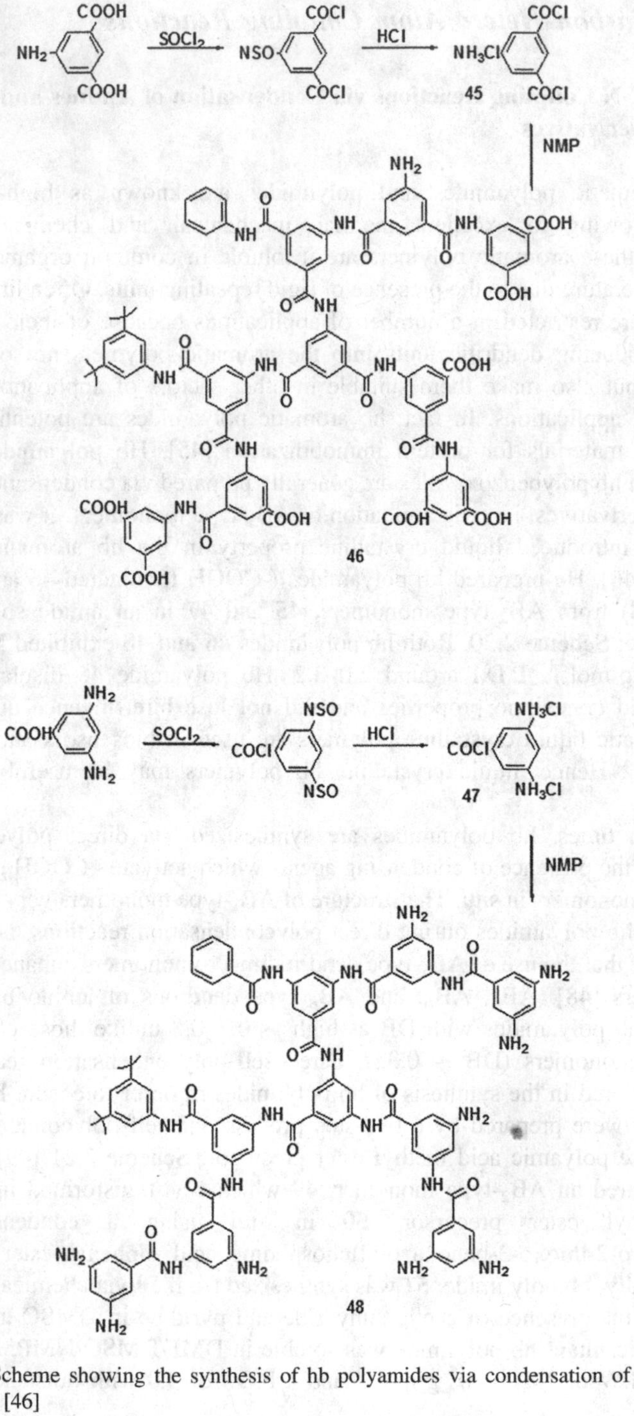

Scheme 2.20 Scheme showing the synthesis of hb polyamides via condensation of amine and acid derivatives [46]

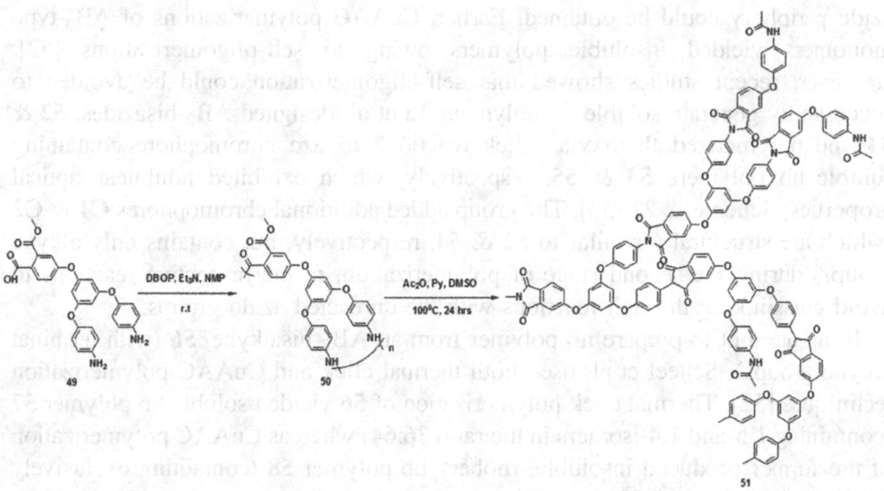

Scheme 2.21 Scheme showing the synthesis of hb polyimides via condensation of amine and acid derivatives [49]

2.2.2.2 Click Chemistry in C-N/C-S Coupling Reactions

[3 + 2] or 1,3-dipolar cycloaddition reactions are often considered as powerful tools in the development of dendritic structures via hetero atom coupling reactions. So far, Cu (I) catalyzed Huisgen 1,3-dipolar cycloaddition between an azide and terminal/internal alkyne derivatives (commonly known as azide–alkyne click reaction or CuAAC reaction), thiol–ene click reaction, and thiol–yne click reaction are extensively used in the synthesis of triazole functionalized hb polymers from various AB_2-type monomers. The term "click chemistry" was first proposed by Sharpless who in his language gave the following criteria for the reaction strategy:

"The reaction must be modular, wide in scope, give very high yields, generate only inoffensive byproducts that can be removed by nonchromatographic methods, and be stereospecific (but not necessarily enantioselective). The required process characteristics include simple reaction conditions (ideally, the process should be insensitive to oxygen and water), readily available starting materials and reagents, the use of no solvent or a solvent that is benign (such as water) or easily removed, and simple product isolation. Purification-if required –must be by nonchromato-graphic methods, such as crystallization or distillation, and the product must be stable under physiological conditions." [50] Sharpless [50] and Tornoe [51] claimed that the reactions to terminal alkynes exclusively yield anti-regioisomer of 1,4-disubstituted [1–3]—triazoles and thus is favored in the polymerization reactions. Click reaction is basically an $A_2 + B_3$ reaction strategy. However, with the advancement of research, many AB_2-type monomers have been designed for the purpose. Depending upon the design of AB_2-type monomers (having both two acetylene and one azide groups or vice versa), hb polytriazoles rich in acetylene or

azide periphery could be obtained. Earlier, CuAAC polymerizations of AB_2-type monomers yielded insoluble polymers owing to self-oligomerizations [52]. However, recent studies showed that self-oligomerization could be avoided to successfully generate soluble hb polymers. Li et al. designed AB_2-bisazides, **52** & **54** and polymerized them via "click reaction" to azo chromophore containing soluble hb polymers **53** & **55**, respectively, which exhibited nonlinear optical properties; Scheme. 2.22 [53]. The group added additional chromophores **C1** & **C2** (which are structurally similar to **52** & **54**, respectively, but contains only alkyne groups) during the second stage of polymerization, in the respective reactions to avoid crosslinking through reactions with the unreacted azido groups.

In an attempt to prepare hb polymer from an AB_2-bisalkyne, **56** (with terminal alkyne groups), Scheel et al. used both thermal click and CuAAC polymerization techniques [52]. Thermal click polymerization of **56** yielded soluble hb polymer **57** (containing 1,5-and 1,4-isomers in the ratio 36:64) whereas CuAAC polymerization of the former produced insoluble, rubbery hb polymer **58** (containing exclusively 1,4-isomer); Scheme 2.23.

In some cases, when 1,5-regioisomer is preferred over 1,4-regioisomer and a faster controlled reaction than CuAAC polymerization is desired, Ru (II) catalyzed click polymerization (RuAAC) is employed [54]. Often click reaction is favored in the synthesis of polymers for biomedical applications. The 1,4-regioisomer of [1–3]—triazole bears a striking resemblance to peptide–amide bond in terms of geometry and thus is compatible with the functional groups present in the biological

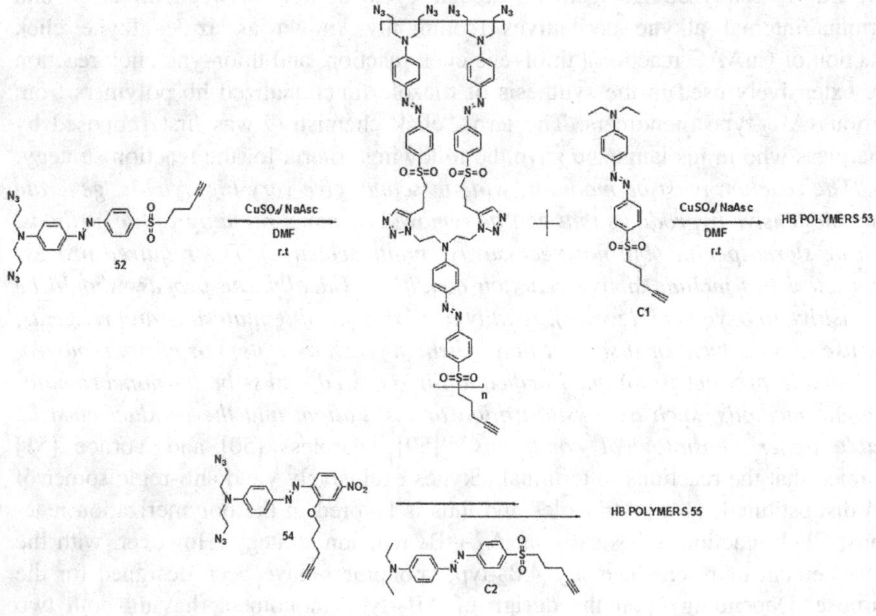

Scheme 2.22 Scheme showing the synthesis of hb polymer from AB_2-bisazides via CuAAC click reaction [53]

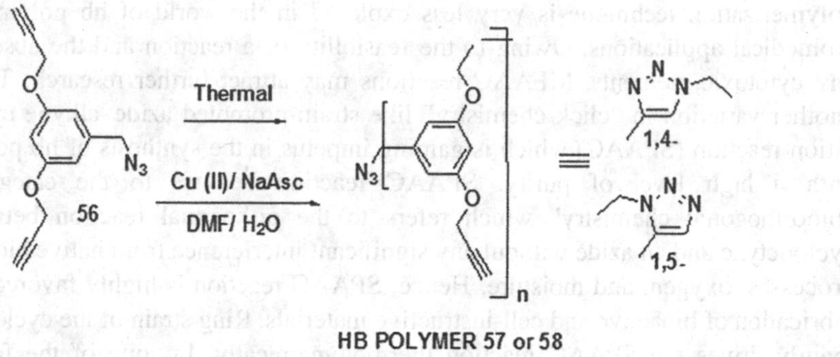

Scheme 2.23 Scheme showing the synthesis of hb polymer from an AB$_2$-bisalkyne via thermal/CuAAC click reaction [52]

macromolecules like proteins, DNA, RNA, etc. In fact, [1–3]-triazole moiety has been utilized to mimic peptide–amide bond to generate dipeptide isotere in ß-strands and α-helical coiled structures [55]. However, there are few drawbacks associated with CuAAC polymerization or rather metal catalyst mediated click polymerization especially if the polymers synthesized are intended for biomedical applications. AB$_x$-type monomers constituting of both alkyne and azide groups are highly reactive and are often difficult to store (even under ambient condition) or are rapidly consumed during the preparation owing to the uncontrolled self-oligomerization; even some materials are explosives. Again, metal catalysts are highly cytotoxic (especially neurotoxic) and the residues are difficult to eradicate from the reaction systems. The presence of trace metal catalysts also affects the physical properties (electronic, optical, etc.) of the synthesized polymers; say for example luminescence of the polymers gets quenched by the metal traps. Even, CuAAC polymerized products are often insoluble which defer the primary necessity of hb polymers. Ru (II) catalysts used in RuAAC polymerization are quite expensive and tedious to prepare. Cu (I) catalysts (either prepared in situ or purchased) are highly sensitive to air and require inert atmosphere for storage which in turn expedite cost. Hence, owing to these serious drawbacks, metal catalyst mediated click polymerization is often discouraged in biomedical applications. In this regard, metal free click polymerization of azides and alkynes, i.e., MFAAC polymerization has emerged as a potential alternative strategy for [1–3]–triazole-based hb polymers syntheses. To be precise, thermally carried out click reactions are not considered as MFAAC reactions because the former does not meet the general criteria for click reactions. In fact, thermally carried out click reactions (which is carried out at quite a high-temperature) are regio-random reactions and yield both 1,4-and 1,5-triazole isomers almost in equal ratios. Generally, alkynes or azides attached to electron withdrawing groups (carboxyl group, ether group, etc.), i.e., propiolates and aroylacetylenes facilitate MFAAC polymerization with high regio-selectivity [56]. The work of Li's group included the synthesis of hb poly (aroxycarbonyltriazole)s via MFAAC polymerization, for the detection of explosives through aggregation induced emission [57]. Surprisingly, MFAAC polymerization technique is very less explored in the world of hb polymers for biomedical applications. Owing to the feasibility of a reaction and the absence of any cytotoxic elements, MFAAC reactions may attract further research. There is another variation in "click chemistry" like strain-promoted azide–alkyne cycloaddition reaction (SPAAC) which is gaining impetus in the synthesis of hb polymers with a high level of purity. SPAAC reaction belongs to the category of "bioorthogonal chemistry" which refers to the orthogonal reaction between a cyclooctyne and an azide without any significant interference from native biological processes, oxygen, and moisture. Hence, SPAAC reaction is highly favored in the fabrication of bioactive and cell-instructive materials. Ring strain of the cyclooctyne mainly drives the SPAAC reaction thermodynamically. In spite of the fact that SPAAC reaction is highly encouraging in labeling biomacromolecules and their use in living cells, owing to the limited commercial availability of the cyclooctyne reagents and tedious synthesis routes, SPAAC is not so much explored yet. With the progress in azide–alkyne click polymerization, recently thiol–ene/yne click polymerization has been much explored either to synthesize hb polymers or to functionalize them. Thiol–ene reaction which occurs between a thiol and an alkene to form an alkyl sulfide, is generally considered as the click reaction (as they are characterized by high thermodynamic driving force and occurs under extremely mild conditions). Thiol–ene reaction proceeds via an anti-Markovnikov addition of a thiol to an alkene and is quite favored in biomedical sciences. Much research has been carried out in the synthesis of hb polymers containing thioether and thioester groups, via free radical or base/nucleophile catalyzed Michael addition type thiol–ene click polymerization. Thiol–ene click polymerization is often followed in the synthesis of hb polycarbosilanes and polycarbosiloxanes and functionalization of other hb polymers. Polycarbosilanes and polycarbosiloxanes are potential antibacterial biocides and thus find too much prospect in biomedical applications including the drug delivery devices [58]. This phenomenon demands extensive research in the exploration of polycarbosilanes and polycarbosiloxanes. Xue et al. successfully synthesized hb organo silicon polymers **60** and **62** via thiol–ene click polymerization under UV light (which is basically a step-growth route involving hydrosilylation between an alkene and a thiol group), of an AB_2-type monomer

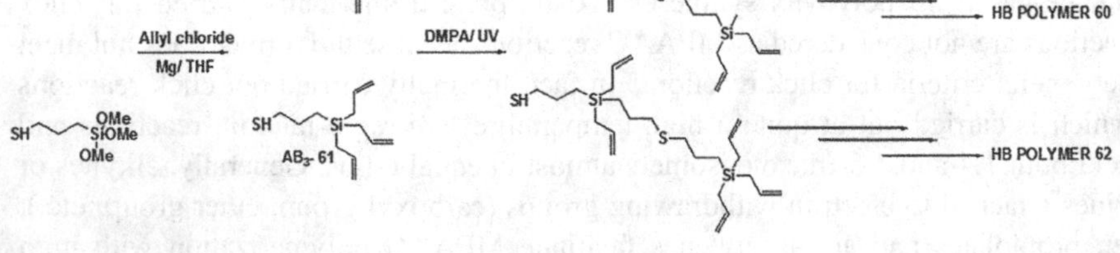

Scheme 2.24 Scheme showing the synthesis of hb organosilicon polymer via step-growth thiol–ene click polymerization [59]

mercaptopropylmethyldiallylsilane, **59** and an AB$_3$-type monomer mercapto-propyltriallyl silane, **61**, respectively, Scheme 2.24 [59]. The hb polymers exhibited an M.W of 3279 g mol^{-1} for **60** and 2963 g mol^{-1} for **62** whereas a DB of 0.6 for **60** and 0.22 for **62**, respectively.

Many times, thiol–ene "click chemistry" is very much useful in the function-alization of polymers for effective biomedical applications. Moreno and coworkers at first developed hb aromatic polycarbosilane hydrophobic cores with allyl/vinyl terminal units, **64** from an AB$_2$-type monomer, **63** via hydrosilylation polymer-ization [60]. The hb polymer **64** exhibited an M.W of 4500 g mol^{-1}, P.D.I of 1.4, and DB of 0.43. Finally, the group carried out different thiol–ene functionalizations on **64** to generate hydrophilic or rather amphiphilic terminal units (anionic or cationic) in the respective hb polymers. The entire scheme for the synthesis of hb polymer is provided in Scheme 2.25.

Similarly, Roy and Ramakrishnan designed an AB$_2$-type monomer, **65** (bearing two allyl benzyl ether groups and an alcohol functionality), allowed self-condensation of **65** under acid-catalyzed melt transetherification to generate hb polymer **66** and then functionalized the peripheral allyl groups using variety of thiols via thiol–ene click reaction; Scheme 2.26 [61].

Another variation, thiol–yne "click chemistry" also holds a strong position in the development of functionalized hb polymers intended for biomedical applications. Thiol–yne click reaction occurs between a thiol and an alkyne in an anti-Markovnikov fashion, to form an alkenyl sulfide. Thiol–yne click polymer-ization proceeds either via thermal or UV initiated route. Konkolewicz and coworkers synthesized a functional hb polymer **68** via photo-initiated step-growth thiol–yne click polymerization from an AB$_2$-type monomer; prop-2-ynyl-3-mercaptopropanoate, **67**, Scheme 2.27 [62]. The group used [2,2-dimethoxy-2-phenylacetophenone]; DMPA as the potential photo initiator. Thiol–yne click polymerization generally yields hb polymers with higher DB than the conventional AB$_2$ polymerization (as the second B group reacts at a faster rate) and offers scope for further functionalization (owing to the presence of many Π bonds in their structures) [6, 63].

The major drawback of thiol–yne click polymerization is the synthesis of suit-able AB$_2$-type monomers containing both free SH and reactive ethynyl groups. These monomers are difficult to prepare and store as they react spontaneously even at low temperature. However, the use of double monomer strategy circumvents the problem of monomer handling for thiol–yne click polymerization. To overcome the problem of monomer handling, Han et al. reported for the first time, synthesis of a reactive AB$_2$-type monomer (containing both thiol and alkyne groups) in situ and its subsequent polymerization to a highly functionalized hb polythioether-yne by following sequential "click chemistry" [64]. In the first step, the group generated an AB$_2$-type monomer via thiol-Michael addition click reaction and in the subsequent step, they polymerized the monomer via thiol–yne click polymerization. The resultant hb polymer exhibited a high DB of 0.6–0.8, high M.W and a broad P.D.I.

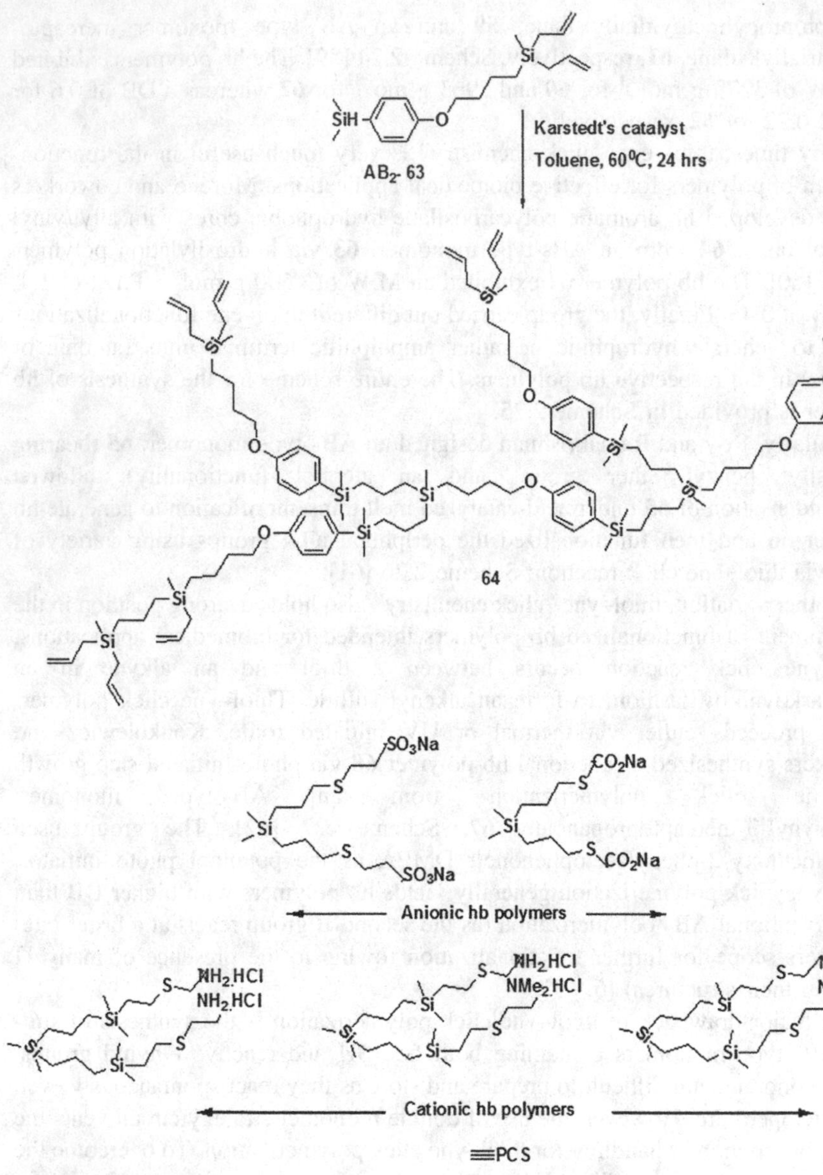

Scheme 2.25 Scheme showing the synthesis of amphiphilic hb polymers via a combination strategy of hydrosilylation polymerization and thiol–ene functionalization [60]

In spite of the fact that thiol-based chemistry is highly favored in the design of biological macromolecules, they are still avoided in many circumstances as thiol functionalized compounds are pungent, sensitive to oxidation, generate harmful reactive oxygen species, and are quite expensive.

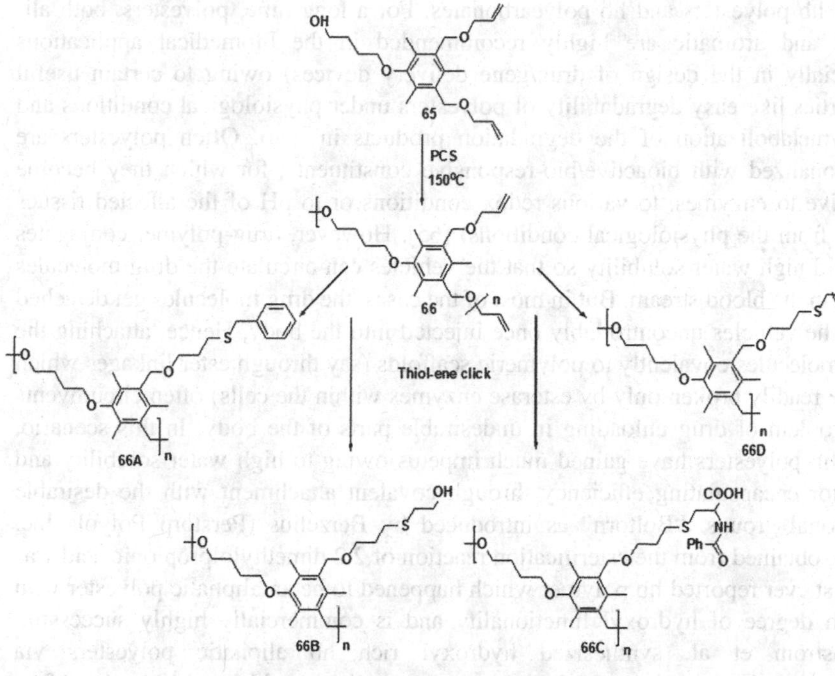

Scheme 2.26 Scheme showing the functionalization of a hb polymer via thiol–ene click reaction [61]

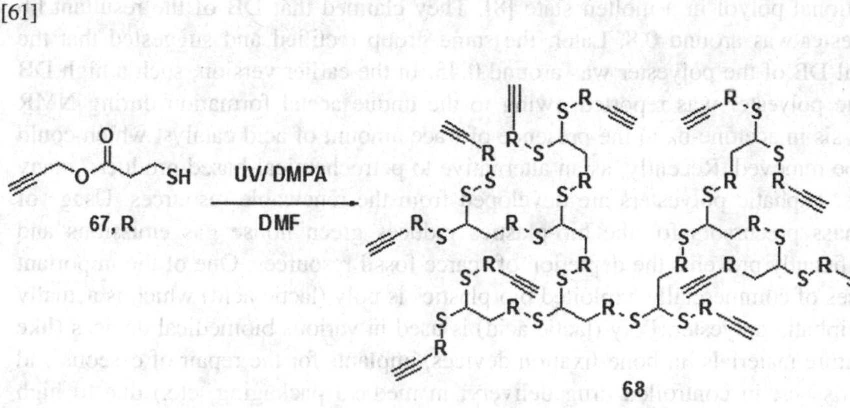

Scheme 2.27 Scheme showing the synthesis of hb polymer via photo-initiated thiol–yne click reaction from an AB_2-type monomer [62]

2.2.2.3   C-O Coupling Reactions

C-O coupling reactions via nucleophilic substitution of alkoxides and phenoxides generally yield hb engineering plastics (e.g., poly (aryl ether)s, poly (ether ketone)s, poly (ether sulfone)s, etc.) which so far are hardly useful in the realm of biomedical applications. However, the esterification of carboxylic derivatives generates very useful hb polyesters and hb polycarbonates. For a long time, polyesters, both aliphatic and aromatic are highly recommended in the biomedical applications (especially in the design of drug/gene delivery devices) owing to certain useful properties like easy degradability of polyesters under physiological conditions and rapid metabolization of the degradation products in vivo. Often polyesters are functionalized with bioactive/bio-responsive constituents; for which they become sensitive to enzymes, to various redox conditions or to pH of the affected tissues (apart from the physiological conditions) [65]. However, drug-polymer conjugates demand high water solubility so that the vehicles can circulate the drug molecules easily in the blood stream. But in most of the cases, the drug molecules get detached from the vehicles uncontrollably once injected into the body. Hence, attaching the drug molecules covalently to polymeric scaffolds (say through ester linkages which can be readily broken only by esterase enzymes within the cells) often circumvents the problem of drug unloading in undesirable parts of the body. In this scenario, thus, hb polyesters have gained much impetus owing to high water solubility and superior encapsulating efficiency through covalent attachment with the desirable functional groups. "Boltorn" as introduced by Berzelius (Perstorp Polyols Inc. USA), obtained from the esterification reaction of 2,2 dimethylolpropionic acid was the first ever reported hb polymer which happened to be an aliphatic polyester with a high degree of hydroxyl functionality and is commercially highly successful. Malmstrom et al. synthesized hydroxyl rich hb aliphatic polyesters via co-condensation reaction of 2,2-bis(methylol)propionic acid; bis-MPA and a four functional polyol in a molten state [8]. They claimed that DB of the resultant hb polyester was around 0.8. Later, the same group rectified and suggested that the actual DB of the polyester was around 0.45. In the earlier version, such a high DB of the polyester was reported owing to the undue acetal formation during NMR analysis in acetone-d_6 in the presence of trace amount of acid catalyst which could not be removed. Recently, as an alternative to petrochemical-based products, many times, aliphatic polyesters are developed from the renewable resources. Usage of biomass precursors for the bio-plastics reduces green house gas emissions and significantly prevents the depletion of scarce fossil resources. One of the important classes of commercially exploited bio-plastics is poly (lactic acid) which is actually an aliphatic polyester. Poly (lactic acid) is used in various biomedical devices (like as suture materials, in bone-fixation devices, implants for the repair of osseous and soft tissues, in controlled drug delivery, in medical packaging, etc.) due to high biocompatibility, biodegradability (can be degraded easily by the hydrolysis of ester linkages without the requirement of any enzymes which otherwise may have caused inflammatory reactions; the hydrolysis of ester linkages even provide spaces for the newly developing tissues) and bio-absorbable properties with low immunogenecity [66]. However, linear poly (lactic acid)s are often difficult to process (due to high crystallinity) and hence are prepared in conjunction with other comonomers [67]. Otherwise, the introduction of branching into the structure also ease the processing of poly (lactic acid) and encourage biomedical applications.

Tasaka et al. synthesized a hb copolymer, **71** of L-lactide (LA), **69** and a metabolically degradable bifunctional monomer; DL-mevalonolactone (ML), **70** via ring opening polymerization, ROP in the presence of Sn(Oct)$_2$ catalyst; Scheme 2.28a [67]. Here, ML (containing a lactone ring and a pendant hydroxyl group) acted as a latent AB$_2$-type comonomer (as the second hydroxyl group remained inactive until the lactone ring was attacked) and also as an initiator for ROP. In another instance, Pitet and his coworkers developed a hb copolymer, **73** of LA, **69** and glycidol, **72** via simultaneous ROP of epoxides and lactides in the presence of Sn(Oct)$_2$ catalyst; Scheme 2.28b [68]. ROP leading to branched architectures is generally favored at a higher polymerization temperature (say around 110–130 °C) and continues for days before desirable products are obtained. To avoid the usage of metallic catalysts, Tsujimoto et al. prepared hb poly (lactic acid) using castor oil (bearing three secondary hydroxyl groups) as the initiator for the ring opening of lactide ring, which finally formed the core of the hb polymer [69]. Apart from poly (lactic acid), ester copolymers of glycerol precursors (obtained from renewable resources) are also considered as important bio-polyesters. Hb polyglycerol or polyglycidol (hb PG) consist of a polyether backbone and peripheral hydroxyl groups at every branch points. Hb PG is generally obtained via oxyanionic ROP of glycidol (a highly reactive hydroxyl epoxide) which acts as a latent cyclic AB$_2$-type monomer (as it releases a second hydroxyl group upon ring opening) [70]. The hydrophilic nature and the presence of free hydroxyl groups in hb PGs suit the design of hydrogels for biomedical applications (drug delivery, tissue engineering, bioconjugation with peptides, protein immobilization, the suppression of protein adsorption to blood-contacting surfaces, etc.) [65, 71]. Moreover, PGs are highly biocompatible, exhibit low cytotoxicity against fibroblast and endothelial cells. For the first time, Sunder's team synthesized hb PGs with controlled M.Ws, narrow P.D.I (1.13–1.47) and reasonable DB (0.53–0.59), via anionic ROP of glycidol and by making use of a fast proton exchange equilibrium (in the presence of a partially deprotonated triol as an alkoxide initiator) [72]. Robinson et al. developed hb aryl polyesters as viscosity improvers (VII) for lubricants via AB$_2$ polycondensation of a monomer containing 12–16 methylene

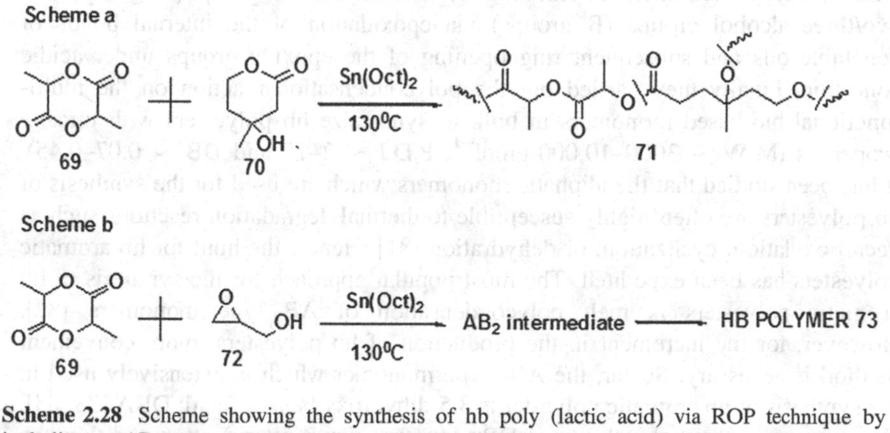

Scheme a

Scheme b

Scheme 2.28 Scheme showing the synthesis of hb poly (lactic acid) via ROP technique by including an AB$_2$-type comonomer and an intermediate, respectively [67, 68]

units in order to ensure good hydrophobicity for solubility in nonpolar medium [73]. Often in high operating temperature window (40–150 °C), lubricants suffer from thinning which adversely reduce application potentiality. Generally, polymers with high M.W (>100 kDa) are used as viscosity modifiers in commercial lubricants [74]. However, linear polymers are prone to degradation under high shear forces. Thus, hb polymers are favored as efficient viscosity modifiers as they are more resistant toward shear degradation [75]. Further, in the subsequent years, various hb copolymers of PGs have also been designed for desirable biomedical applications. Lee and his team made an approach to develop hb double hydrophilic block copolymer of poly (ethylene oxide)-hb-polyglycerol as an efficient drug delivery system with high loading capacity and controlled release properties [76]. The hb copolymer was capable of forming a self assembled micellar structure on conjugation with doxorubicin (a popular hydrophobic anticancer drug) when linked through pH sensitive hydrazone bonds. Following the protocol of Sunder and his coworkers, Garamus et al. prepared amphiphilic hb poly (glycerol ester)s with varying degrees of esterification, by partial esterification of a hb PG with palmitoyl chloride and studied their solution properties in different solvents, using SANS studies [72, 77]. Parzuchowski and his coworkers also synthesized highly functionalized hb polyesters from glycerol-based AB_2-type monomers, ethyl{3-[2hydroxy-1-(hydroxymethyl)ethoxy]propyl}thioacetate via polycondensation in the presence of different catalysts [78]. Other biocompatible hb aliphatic polyesters include hb polycarbonates (hb PCs) and hb polyphosphates which have been explored in the last few years for various biomedical applications. Parzuchowski and his team successfully developed an AB_2-type monomer, 5-{3-[(2-hydroxyethyl)thio)]propoxy}-1,3-dioxan-2-one from glycerol and subsequently polymerized the monomer to biocompatible and biodegradable hb PCs via ROP technique [79]. However, most of the hb PCs and hb polyphosphates are generated via $A_2 + B_3$ strategy and thus detailed under the Sect. 2.3. In a recent work, Testud and his team successfully developed hb polyesters via polycondensation of fatty acid-based AB_x-type monomers [80]. The group designed four different types of AB_2/AB_3-type monomers constituting of a methyl ester group (A group) and two/three alcohol groups (B groups) via epoxidation of the internal bonds of vegetable oils and subsequent ring opening of the epoxide groups under acidic condition. Finally they carried out the polycondensation reaction on the multifunctional bio-based monomers in bulk to synthesize hb polyesters with tunable properties (M.W $\sim$ 3000–10,000 $gmol^{-1}$, P.D.I $\sim$ 2–15 and DB $\sim$ 0.07–0.45). It has been studied that the aliphatic monomers which are used for the synthesis of hb polyesters are often highly susceptible to thermal degradation reactions such as decarboxylation, cyclization, or dehydration [81]. Hence, the hunt for hb aromatic polyesters has been expedited. The most popular approach for the synthesis of hb aromatic polyesters is melt polycondensation of AB_2-type monomers [82]. However, for the increment in the production of hb polyesters, more convenient method is necessary. So far, the AB_2-type monomer which is extensively used in the synthesis of hb aromatic polyester is 3,5-dihydroxy benzoic acid; DBA [83, 84]. However, poor thermal stability of DBA restricts direct esterification and thus the hydroxy groups are often chemically modified by acetylation or trimethylsilylation prior to polycondensation reactions. In the earliest known work, Kricheldorf and his team used 3,5-bis(trimethylsiloxy)benzoyl chloride as an AB_2-type monomer and condensed it with 3-(trimethylsiloxy)benzoyl chloride in bulk (at around 250–300 °C) to synthesize hb poly (3-hydroxybenzoate) which is an aromatic polyester [85]. They isolated the hydroxyl terminated hb polyesters from the synthesized hb polymers by adding methanol which hydrolyzed trimethylsiloxy groups. Similarly, Turner and his group used 3,5-diacetoxybenzoic acid as an AB_2-type monomer and condensed it via acidolysis reaction at 250 °C in bulk to synthesize a hb aromatic polyester [86]. The resultant hb polyester was soluble and exhibited a M.W greater than 10^6 g mol^{-1}. Generally, the polycondensation reaction of 3,5-diacetoxybenzoic acid requires higher polymerization temperature than that required for the polycondensation reaction of 3,5-bis(trimethylsiloxy)benzoyl chloride in order to obtain a polymer with high M.W. Fomine and his group reported the synthesis of a coumarin (which has medical approval in pharmaceutical chemistry) containing hb aromatic polyester via the esterification reaction of an AB_2-type monomer [87]. Kricheldorf and his group reported the polycondensation reaction of an AB_3-type monomer (triacetylated gallic acid) in bulk to generate a hb aromatic polyester [88]. Qie et al. reported the synthesis of carboxylic groups terminated aryl-alkyl hb polyester via melt polycondensation of an AB_2-type monomer, 5-hydroxyethoxyisophthalic acid [89]. The resultant hb poly (5-hydroxyethoxyisophthalic acid) being a polycation, was used to generate self-assembly films with the assistance of a polyanion, poly (diallyldimethylammonium chloride) via layer by layer technique. Often, to replace the robust condition of melt/bulk polycondensation process, one-pot solution polycondensation of AB_2-type monomer is favored in the synthesis of hb polyesters. Moreover, in the melt polycondensation process, if even a trace amount of impurities is present in the monomer then insoluble polymers are formed [90]. In this regard, Erber et al. showed the formation of hb aromatic polyesters with phenol terminal groups from an AB_2-type monomer, 3,5-dihydroxybenzoic acid via solution polycondensation reaction, in the presence of 4-(dimethylamino) pyridinium 4-tosylate as the catalyst [91]. The resultant hb aromatic polyester exhibited a DB around 0.6 which was almost similar to those hb aromatic polyesters obtained via melt polycondensation of 3,5-bis(trimethylsiloxy) benzoyl chloride. Owing to the broad utility of the hb aromatic polyesters (in coatings, paints, adhesives, etc.), extensive researches are carried out to develop new interesting properties and even some of them have been successfully explored for commercialization. Generally, hb aromatic polyesters are used in vitro biomedical applications due to the question of biocompatibility of all grades.

2.2.2.4 Enzyme-Catalyzed Polymerization

These days, there is an urge for the development of nontoxic and environment friendly catalysts for the polymer syntheses. As an alternative to the existing metal catalysts used in polymerization, isolated enzymes have attracted much attention.

Enzymes exhibit high catalytic activities, offer good enantio/regio/chemoselectivity, follow mild reaction conditions, have the ability to be used in bulk reaction media (without the use of organic solvents), are biodegradable, recyclable and maintain good biocompatibility [92]. Hence, enzyme-catalyzed polymerization is often highly favored in the synthesis of polymers specially intended for biomedical materials, drug/gene delivery vehicles and other pharmaceutical materials. So far, enzyme-catalyzed polymerization has successfully yielded many polymers like polyesters, polycarbonates, poly (amino acid)s, polyaromatics, etc. Popular commercially exploited enzymes, widely used in the polymerization have been listed by Uyama and Kobayashi which include oxidoreductases (for polyphenols, polyanilines), transferases (for polysaccharides, polyesters), hydrolases (for polysaccharides, polyesters, polycarbonates, etc.), lyases, ligases, etc. [93, 94]. Initially, enzyme-catalyzed polymerization of few selective monomers like ξ-caprolactone (CL), δ-valerolactone (VL), and γ-butyrolactone (BL) yielded polymers with low M.W and that too after many hours of reaction [95]. Hence, to circumvent the problems associated with the earlier version of enzyme-catalyzed polymerization, these days the polymerization is carried out using immobilized enzyme catalyst subtrates [92]. With the advancement in research, enzyme-catalyzed polymerization has also been explored in the synthesis of hb polymers which are more likely to suit biomedical applications. For the first time, Skaria et al. synthesized a series of hb copolyesters via a combination of ring opening AB polymerization (of ξ-caprolactone) and AB_2 polycondensation (of 2,2-bis(hydroxymethyl)butyric acid, BHB), catalyzed by immobilized Lipase B (isolated from Candida antarctica) [95]. They were able to maintain $0 < DB < 0.33$ for the different hb copolyesters just by controlling the comonomer ratio in the feed. Lopez-Luna and coworkers reported the synthesis of hb poly (VL-co-BHB) and poly (CL-co-BHB) via immobilized Lipase B catalyzed ring opening of the respective L-lactide in the presence of an AB_2-type comonomer BHB core [96]. They carried out the enzyme-catalyzed ROP in 1,1,1,2-tetrafluroethane as a green, benign, polar, but hydrophobic solvent. However, the resultant hb copolyesters were semi crystalline and exhibited low DB (0.02–0.09) which might be due to the limited solubility of BHB in 1,1,1,2-tetrafluoroethane (~ 10 wt%). Later the same group improved DB of the resultant hb polylactones by carrying out the enzyme-catalyzed ROP of L-lactide (CL/VL) and an AB_2-type core monomer BHB, in an ionic liquid (IL); 1-butyl-3-methylimidazolium hexafluorophosphate which is again considered as an environmentally benign solvent [97]. In a recent work, Xu et al. synthesized a series of hb poly (amine-ester)s with a value of DB (>0.8), via immobilized Lipase B catalyzed polycondensation of triethanolamine and diesters [98]. The resultant hb poly (amine-ester)s were biodegradable, exhibited low cytotoxicity/good biocompatibility/micellization ability and thus were suitable for loading/carrying drugs. There is immense scope in enzyme-catalyzed polymerization as many new enzymes from different sources are coming up commercially at acceptable prices. Hence, further research in the realm of enzyme-catalyzed polymerization would definitely develop new hb polymers with huge potentiality in biomedical applications.

2.3 Hyperbranched Polymers from $A_2 + B_3$ Monomer Pairs and Other Couple Monomer Methodologies

Apart from SMM (like AB_x polymerization), the application of double monomer methodology; DMM in the synthesis of hb polymers is also highly acknowledged as an alternate strategy. DMM is further classified into $A_2 + B_3$ methodology and couple monomer methodology; CMM, depending upon the selection of monomer pairs and reaction pathways. Polymerization of the functionally symmetrical monomer pairs like A_2/B_4, A_3/B_3, A_2/B_3, etc., yields soluble hb polymers. Among these, $A_2 + B_3$ DMM (via condensation or addition mechanism) has received significant encouragement in the synthesis of hb polymers. A_2 and B_3 being separate entities are relatively easy to synthesize as compared to AB_x monomers and thus facilitate commercialization. There is enough scope in expanding $A_2 + B_3$ strategy for biomedical applications owing to the existence/development of the vast choices of the monomer pairs. $A_2 + B_3$ strategy can be followed either in polycondensation or in self condensing vinyl polymerization. However, $A_2 + B_3$ polymerization often occur in an uncontrolled fashion and there is always a high risk of gelation and intra molecular cyclization reactions (especially when the molar feed ratio of A_2: $B_3 > 0.9$, at high monomer concentrations and at high conversions) [99]. In an $A_2 + B_3$ approach, AB_2-type intermediates are formed at the initial stage of polymerization. In the subsequent reaction steps, AB_2-type intermediates further reacts with unreacted A_2 and B_3 monomers and produces A_xB_y species (where $x \geq 2$ and $y \geq 2$). These A_xB_y being highly reactive species encourage intra molecular cyclization and crosslinking (after a certain conversion of the functional groups). In general, $A_2 + B_3$ strategy develops branched polymers with cyclic building blocks and/or a mixture of branched and cyclic polymers. In fact, earlier $A_2 + B_3$ strategy was solely used for the synthesis of crosslinked polymers. Aharoni et al. established a series of successful reactions between aromatic diamines (A_2) and aromatic dicarboxylic acids (B_3) but the resultant polymers were crosslinked networks [100, 101]. It was only after the pioneer work of Jikei and Kakimoto, it was established that when equimolar amounts of A_2 and B_3 monomers are used, soluble hb polymers could be synthesized [102]. The group prepared hb aromatic polyamides via condensation reactions between aromatic diamines (A_2) and trimesic acid (B_3) in the presence of condensing agents at 80 °C for 3 h. In this work, they maintained a low monomer concentration of monomers; 0.21 mol L^{-1} (3.3 wt%) to avoid gelation. They proposed that if the first condensation of A_2 and B_3 is faster than the following propagation, then an AB_2-type intermediate is formed which subsequently undergoes polycondensation and thus prevents gelation. In fact, they compared the structural features of hb aromatic polyamides obtained from $A_2 + B_3$ polymerization and AB_2 polymerization. To their surprise, they observed that the hb aromatic polyamides obtained from $A_2 + B_3$ strategy exhibited higher number of the dendritic units as compared to the terminal units whereas those obtained from AB_2 polymerization exhibited equal number of the dendritic and the terminal units. Following this work, numerous patents and papers have been published which were mostly based on the poly-condensation between glycerols and dicarboxylic acids/or cyclic anhydrides. So far, $A_2 + B_3$ strategy has been employed successfully to synthesize some important classes of hb polymers like hb polyamides, hb polyimides, hb polyesters, hb polyethers, hb polycarbonates, hb polyphosphates and hb polyurethanes; some of them are undoubtedly potential biomaterials. Fang et al. used $A_2 + B_3$ strategy to develop hb aromatic polyimides [103]. In this work, they used a series of dian-hydrides (A_2) and a triamine; tris(4-aminophenyl)-amine (B_3) to prepare the hb polyamic acid precursors which were subsequently transformed to the respective hb aromatic polyimides via thermal or chemical imidization. Here, the order of monomer addition played a significant role in the development of hb aromatic polyimides either rich in terminal amines or anhydrides and it hardly affects DB. Often it is difficult to avoid gelation in an ideal $A_2 + B_3$ polymerization. Deviation from the ideal $A_2 + B_3$ strategy, i.e., by using a condensing agent, it is possible to prevent gelation significantly. Hao et al. followed a non ideal $A_2 + B_3$ strategy to prepare soluble hb polyimides from 1,4-phenylenediamine (A_2) and tri (phthalic acid methyl ester) (B_3) in the presence of diphenyl (2,3-dihydro-2-thioxo-3-benzoxazoyl) phosphonate condensing agent [104]. These hb polymers exhib-ited DB of 0.52–0.56 and inherent viscosities of 0.17–0.97. Unal and Long developed a hb poly (ether ester) via cyclization free melt condensation of A_2 oligomers and B_3 monomers [105]. In this novel work, they condensed oligomers of poly (propylene glycol) and trimethyl 1,3,5-benzenetricarboxylate in the pres-ence of titanium tetraisopropoxide and stopped the reaction prior to gelation. Both hb PCs and hb polyphosphates are highly useful as functional materials and bio-medicine, such as antibacterial/antifouling materials, in protein purification/detection/immobilization/delivery, in drug/gene delivery, in tissue engineering, in bioimaging, etc. Scheel et al. synthesized thermo-labile hb PCs via $A_2 + B_3$ route employing bis (carbonylimidazolide) and triethanolamine [106]. Such hb PCs may find usefulness in the preparation of nanoporous materials. In another work, Miyasaka and coworkers synthesized hb PCs via $A_2 + B_3$ polycondensation of di-tert-butyl tricarbonate (A_2) and 1,1,1-tris(4-hydroxyphenyl)ethane (B_3) [107]. The resultant hb PCs exhibited DB around 0.5–0.7. Apart from being biocompat-ible and biodegradable, polyphosphoesters are structurally similar to nucleic acids and teichoic acids. Under physiological conditions, polyphosphoesters easily degrade into harmless, low M.W materials either through hydrolysis or enzymatic degradation of the phosphate bonds. In an early work, Wang and Shi developed a reactive flame retardant hb polyphosphoester via $A_2 + B_3$ polycondensation of bisphenol-A (A_2) and phosphoryl trichloride (B_3) at 100 °C [108]. Following this work, many research has been focused on the synthesis of hb polyphosphates and functionalized hb polyphosphates which is elaborately detailed in the review of Lie et al. [109]. There are also significant works on the synthesis of soluble hb polymers via $A_2 + B_3$ CuAAC polymerization. Xie et al. prepared hb polytriazoles via $A_2 + B_3$ CuAAC polymerization of 4-N,N′-bis(2-azidoethyl)amino-4′-nitroazo-benzene (A_2) and 1,3,5-tris(alkynyloxy)benzene (B_3) in a one pot at room tem-perature [110]. In another work, Qin et al. developed soluble, regio-regular hb poly

(1,2,3-triazole)s via A$_2$ + B$_3$ CuAAC polymerization [111]. The resultant hb polytriazoles exhibited DB around 0.9 and quite high M.W. Chen et al. developed reduction cleavable disulfide bonds containing hb poly (ester triazole)s via A$_2$ + B$_3$ CuAAC polymerization of dipropargyl 3,3'-dithiobispropionate (A$_2$) and tris(hydroxymethyl)ethane tri(4-azido butanoate) (B$_3$) [112]. The hb poly (ester triazole)s exhibited M.W of 2.04×10^4 g mol^{-1} and P.D.I around 1.57–2.17. These hb polymers are highly recommended as stimuli responsive anticancer drug nanocarriers; except the fact that catalyst traces have to be eradicated. Often A$_2$ + B$_3$ strategy facilitate the synthesis of hb polymers with exceptional desirable properties, from uncommon monomer couple pairs. Kanai et al. designed hb poly (cyanurateamine) and hb poly (triacrylatetrimine) with good antimicrobial activities, from a pair of multifunctional monomers, via A$_2$ + B$_3$ Michael addition type polymerization [113]. In this work, they used diethylenetriamine and 2,4,6-triallylcyanurate/trimethylolpropane triacrylate as A$_2$ and B$_3$ monomer pairs, respectively. From the growing demand for the development of A$_2$ + B$_3$ strategy as an alternative to AB$_2$ polymerization, it is obvious that A$_2$ + B$_3$ polymerization has a very good prospect in future. Hence, to encourage the establishment of A$_2$ + B$_3$ strategy for commercialization, gelation must to be prevented. Some ways of preventing gelation include the formation of reactive AB$_2$-type intermediates (as gelation is easier to prevent in AB$_2$ polymerization), partial functionalization of peripheral groups, stopping the polymerization through precipitation/end capping prior to the critical point of gelation, usage of suitable condensing agents, intense stirring, keeping low monomer concentration, and through other few methods as described later under the Sect. 2.4.

Although there are numerous methods to prevent gelation in A$_2$ + B$_3$ polymerization, yet most of them are not effective to the mark. Hence, as an alternative to A$_2$ + B$_3$ strategy, a new genre of synthesis strategy was developed. This new synthesis strategy is based on the in situ formation of AB$_x$ intermediates (during the initial stage of polymerization) from a specific pair of monomers with different reactivities of functional groups in each. These monomer pairs are basically functionally asymmetric in nature. This approach is analogous to nonideal A$_2$ + B$_3$ strategy and is popularly known as couple monomer methodology; CMM. Owing to the prevention of gelation, CMM is highly sought after for the scale up of hb polymers. Depending upon the reactivity of the monomer pairs, different categories of CMM have been reported [114]. For a monomer pair of AA' and B'B$_2$, in one of the instance when A' is more reactive than A then CMM is named as AA' + B$_3$ strategy. In another instance, when B' is more reactive than B then CMM is named as A$_2$ + B'B$_2$ strategy. Finally, when both A' and B' are differently reactive than A and B, respectively, then CMM is named as AC + DB$_2$ strategy. In all the categories of CMM, AB$_2$-type intermediates are formed in situ which eventually react to form hb polymers without any crosslinked structures. In CMM, M.W and the type of terminal functional groups on the hb polymers can be controlled by maintaining the molar feed ratio of the monomer pairs. When the molar feed ratio of AA' to B'B$_2$ is lower than 1:1 then the resultant hb polymers exhibit low M.W and are rich in B groups. Whereas when AA':B'B$_2$ molar feed ratio is greater than 2:1

then the resultant hb polymers have high M.W and are rich in A groups. It is only when $AA':B'B_2$ molar feed ratio is between 1:1 and 2:1, the hb polymers are equally rich in A and B groups with moderate M.W. So far, CMM has been successfully established in the generation of hb poly (sulfone amine)s, hb poly (ester amine)s, hb poly (urea urethane)s, hb poly (amide amine)s, hb polyesters, and hb poly (ester amide)s [114, 115].

2.4   Drawbacks of Hyperbranched Polymerization Techniques and Possible Remedies

From the vast studies of AB_x polymerization or rather AB_2 polymerization for the synthesis of hb polymers, it is clear that AB_2 polymerization is a very important category of reaction. However, in an AB_2 type batch polycondensation reaction, often there is hardly any control over the polymerization conditions owing to the complementary reactivity of A and two B functional groups which lowers M.W, broadens P.D.I, and DB is hardly beyond 0.5 for the resultant polymers [90]. Even yields of the polycondensed hb polymers in a step-growth reaction decrease owing to internal cyclization reactions, crosslinking reactions, and different reactivities of the similar B functional groups. Ideally, an AB_2-type hb polymer should resemble the topology of a perfect dendrimer as two branches are expected to develop in a regular fashion, from each monomer. Although the two B groups of a terminal unit are equal in reactivity yet their ability to react depends strongly on the kinetic factors prevailing during the one-pot step-growth process. Once one of the B groups get reacted, a proximal steric hindrance is generated which generally prevents the other B group to react and thus the probability of development of branching units decrease (as observed in most of the reactions). Again, the regular dendritic growth of an AB_2-type monomer gets impeded due to fast depletion of the monomer at an early stage of the reaction and thereby the polymerization continues through coupling of the sterically hindered oligomeric units. Thus, these limitations of AB_2 polymerization hinder the development of hb polymers as advanced soft materials in the commercial world. Lot of efforts has been made to overcome the shortcomings of AB_2 polymerization in order to add prosperity to the subject. In this regard, Shi et al. made a detailed and comprehensive study on the ongoing development of new reaction strategies for the synthesis of hb polymers with controlled topology and properties [116]. They claimed that there are three categories of reaction strategies which may be followed for any type of polymerization (for step growth, chain growth, and other reaction types) to improve the structure and properties (controlled M.W, narrow P.D.I and high DB) of the hb polymers (Scheme 2.29). These include-

(I)   Slow addition of monomers to multifunctional core or chain terminator (a semi batch process). From the pioneer work of Malstorm et al. [8], Feast et al. [117] and Bharathi et al. [118], a concept of copolymerization of an

Approach 1: slow addition of monomer to core

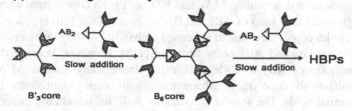

Approach 2: high reactivity of core molecule

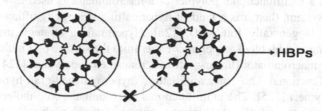

Approach 3: confined space

Scheme 2.29 Schematic representation of three different approaches to decrease P.D.I of the hb polymers. Reprinted (adapted) with permission from Shi et al. [116]. Copyright (2015) American chemical society

AB_2-type monomer in the presence of a multifunctional core monomer, B_f was developed to control M.W and P.D.I of the hb copolymers significantly. During any polymerization reaction, probability of the addition of hb species to the other species (monomers, oligomers, etc.) is proportional to the number of functional groups and the number of functional groups is again proportional to DP. Thus, larger molecules grow faster than the smaller ones which in turn actually broadens M.W.D of the resultant hb copolymers [119]. Due to slow monomer addition, a very slow monomer concentration is maintained throughout, in the reaction mixture. Thus, the monomer almost exclusively reacts with the growing polyfunctional macromolecules (with complete exclusion of monomer–monomer reaction) which results in high M.W, narrow P.D.I, and high DB [120, 121]. Moreover, the B_f molecule acts a chain terminator core which also restricts the undesirable side reactions significantly. In an ideal situation, the conversion of A groups should be 100% at all times of the reaction, i.e., the previously added monomer has to be consumed completely before a new monomer is being added, which is however not the case. But in reality, this approach is favorable when the rate of polymerization is reasonably fast. However, it is quite challenging to maintain the rate of monomer addition to be reasonably slow but constant during the polymerization reaction. The addition of a B_f multifunctional core molecule to an AB_2 system also makes the final hb copolymer to resemble the dendrimer like hierarchical architecture. In a random polymerization process, as most of the terminal units disperse at the periphery, the introduction of a core adds a focal point in the hb copolymer [122]. The molar feed ratio of the AB_2-type monomer and the B_f molecule is a highly critical parameter in determining M.W and P.D.I of the hb copolymers. Generally, a high molar feed ratio of AB_2 and B_f (say greater than 100) broadens M.W.D of the hb copolymers due to increased probability of the monomer–monomer reaction, i.e., self AB_2 polymerization [118]. However, keeping the molar feed ratio of AB_2 to B_f less than 100 dramatically reduces M.W of the resultant hb copolymers which may limits their applications in many potential fields. The second comonomer multifunctional core molecule may be added to the AB_2 system, either as a small molecule or as a polymer. When a multifunctional polymer or a macroinitiator is used as a core in an AB_2 system, then it is a kind of hypergrafting strategy (grafting-from technique is generally followed) [123]. Hypergrafting strategy may involve grafting of a hb block copolymer either from the surface of a multifunctional linear macroinitiator (linear-graft-hyperbranched copolymer) [124] or from a multifunctional hb macroinitiator (hyperbranched-graft-hyperbranched copolymer) [125]. The multifunctional macroinitiator core molecule is generally composed of polyglycerols and poly (ethylene glycol)s. It has been found that hypergrafted hb copolymer products being core–shell type polymers are often highly useful in biomedical transport applications or for applications in bio-conjugations. In spite of using a multifunctional core molecule in the polymerization reaction mixture, often hb copolymers exhibit bimodal M.W.D; a narrow, high M.W peak, and a broad, long tail on the lower M.W end. Bharathi and Moore dramatically improved the GPC chromatograms of the hb copolymers prepared from an AB_2-type monomer and a B_f comonomer by carrying out the polymerization reaction on an insoluble solid (polymer, peptide, etc.) support [126]. They speculated that due to the limitations in the spaces that exists when a hb polymer attaches to a solid support, a high M.W could be maintained during the hb polymerizations. The solid support also inhibited any undesirable intramolecular cyclizations between the A and the B groups. In this novel work, they tethered a B_f-type monomer core to an insoluble solid support suspension through a triazene linkage and then subsequently polymerized an AB_2-type monomer in the presence of the core support suspension, Pd (II) catalyst and piperidine. The resultant hb copolymers (isolated from the solid support) exhibited highly controlled features like M.W of around $5\text{--}25 \times 10^3$ g mol^{-1} and P.D.I of around 1.1–1.5. The only problem associated with the technique of slow addition of the AB_2-type monomer to a multifunctional core (either supported or unsupported) is the happening of self-polymerization of the AB_2-type monomer which generally results in low yield of the desirable hb copolymers.

(II) Maintaining different reactivities of the multifunctional core and the mono-mer in a real polymerization process, although a difference in reactivity between the AB_2-type monomer and the B_f molecule is always maintained but by maintaining a wide difference in their reactivities, further improve-ment in the GPC chromatogram of the resultant hb copolymers is possible [127]. This approach is analogous to slow monomer addition in the sense that randomness of the polymerization reaction is significantly reduced. Due to wide difference in the reactivity ratio between AB_2 and B_f (B_f is preferably more reactive than the B groups of the AB_2-type monomer due to higher degree of functionality), B_f reacts with AB_2 at a faster rate and delays undesirable homopolymerization of AB_2. Bernal et al. prepared hb poly (arylene ether phosphine oxide)s in the presence of a series of B_f molecules with varied reactivities and observed P.D.I as low as 1.25 [128]. Following this work, further improvement in the GPC chromatogram was procured by Roy et al. where they synthesized hb polyethers in the presence of highly active B_3 core than the AB_2-type monomer [129]. They observed on increasing the molar fraction of B_3 to AB_2, decrease in M.W (as low as 5400 g mol^{-1}) and P.D.I (<1.5) occurred. Hence, in order to obtain hb copolymers with narrow P.D.I but high M.W and high DB in the true sense, more research has to be done to develop new pathbreaking strategies. These days, activation of the functional groups on the core polymer also signifi-cantly narrows M.W.D of the hb copolymers. On using the B_f core molecule with higher reactivity than the AB_2-type monomer, a faster consumption of the core molecules than the AB_2-type monomer occurs in the initial stage. However, after complete reaction of the B with the A groups, the surface B groups on the produced 1st generation hb copolymer become equally reactive to that of the B groups on the monomer. This phenomenon significantly reduces the effect of faster reaction of the B_f core than the monomer. Hence, a consecutive activation of the B groups on the 1st generation hb copolymer during the polymerization reaction is necessary. Suzuki et al. introduced the concept of multibranching polymerization (MBP) which generates a den-dritic polymer with the initiator as the core [130]. MBP proceeds via mul-tiplication of the propagating ends at every step of the propagation. To carry out a successful MBP reaction, a judicious choice of monomers and an adequate feed ratio between the monomers has to be maintained. In a later work, Suzuki et al. designed hb dendritic polyamine consisting of primary, secondary (a nonbranching point) and tertiary (a branching point) amino groups via Pd catalyzed decarboxylative ROP of a cyclic carbamate mono-mer in the presence of benzylamine initiator [131]. This type of MBP involves a monomer with a leaving group at the allylic position which is activated from time to time by the oxidative addition of Pd to form a Π-allylpalladium complex intermediate. The Π-allylpalladium intermediate is further nucleophilically attacked by an initiator fragment and a propagating end. Thus, here free amine initiator (benzylamine) or rather the growing polymer chains reacted with the monomer but the monomer could not undergo self-polymerization; thereby narrows M.W.D of the resultant hb copolymers. The hb copolymers exhibited DB around 0.6–0.8 and P.D. I < 1.5. In another attempt, Ohta and coworkers synthesized hb copolymers by exploiting the change of substituent reactivity effects between the monomer and the B_f core as well as the growing polymer [132]. Generally, reactivity of the initially identical groups changes after one of them has reacted. In their study, they generated an AB_2-type monomer anion with an amide anion (as A group) and two ethyl ester groups (as B groups) where the amide anion deactivated both the ester groups which subsequently prevented AB_2 self-polymerization (through +I effect) and encouraged reaction of the AB_2-type monomer anion exclusively with the initiator core molecule (constituting of two reactive ester groups). In the following step, the AB_2-type monomer anion further reacts with the ester groups of the growing polymer chain which had been activated through the formation of amide linkage. The resultant hb copolymers exhibited P.D.I < 1.13 but DB was only 0.5 (due to equal reactivity of the terminal B groups on the polymer). Unfortunately, this approach is applicable only to certain classes of polymers (as the choice of monomer-core pair is critical) which restrict its exploration.

(III) By carrying out the hb polymerizations in confined spaces- When a polymerization reaction (especially in FRP) is carried out in a homogeneous medium, variety of random reactions occur due to unconfined nature of the process (i.e., infinite space). Undoubtedly, such random reactions adversely affect the structure of the polymers with complex architectures where otherwise control in the reaction strategy is necessary. On the contrary, when a polymerization is carried out in a heterogeneous, dispersed medium (say in microemulsion), it is possible to synthesize polymers with controlled and complex architectures. A well dispersed medium or rather the presence of discreet micelles ensures compartmentalization which is basically segregation and/or confinement of the reactants within the discrete polymerizing particles. Such confinement of the reactant molecules dramatically reduce polymer–polymer inter particle reactions and thus narrows M.W.D of the resultant hb polymers. Each discrete polymerizing sites acts either as a microreactor or a nanoreactor with a well defined boundary. Each reactor is highly localized and independent and there is hardly any mass transfer between the adjoining reactors. This synthesis approach is mainly popular in the controlled radical polymerization technique. So far, the attempt to control the hierarchical structures of the hb copolymers with well defined properties has been mostly successful with this approach.

2.5 Conclusion

This chapter has covered some of the notable advancements in the synthesis of hb polymers via step-growth and chain-growth reactions since their birth. Typical approaches for the synthesis of hb polymers like AB_x polycondensation, $A_2 + B_3$ polymerization and other categories which are quite widely followed worldwide is discussed in length. In each section, a brief detail has been provided how different methods can be used to produce different important classes of hb biomaterials. This chapter also focuses on the drawbacks associated with the different preparatory methods for the hb polymers and how various researches have come up with suitable remedies.

References

1. Gao C, Yan D, Frey H (2011) Promising dendritic materials: an introduction to hyperbranched polymers. In: Hyperbranched polymers, John Wiley & Sons, Inc.: pp 1–26
2. Voit B (2000) New developments in hyperbranched polymers. J Polym Sci A Polym Chem 38(14):2505–2525
3. Flory PJ (1953) Principles of polymer chemistry. 1st ed. Cornell University Press, Ithaca, United States
4. Flory PJ (1941) Molecular size distribution in three dimensional polymers. I. Gelation1. J Am Chem Soc 63(11):3083–3090
5. Flory PJ (1952) Molecular size distribution in three dimensional polymers. VI. Branched polymers containing A-R-Bf-1 type units. J Am Chem Soc 74(11):2718–2723
6. Hölter D, Burgath A, Frey H (1997) Degree of branching in hyperbranched polymers. Acta Polym 48(1–2):30–35
7. Beginn U, Drohmann C, Moller M (1997) Conversion dependence of the branching density for the polycondensation of AB n monomers. Macromolecules 30(14):4112–4116
8. Malmstrom E, Johansson M, Hult A (1995) Hyperbranched aliphatic polyesters. Macromolecules 28(5):1698–1703
9. Hawker CJ, Lee R, Frechet JMJ (1991) One-step synthesis of hyperbranched dendritic polyesters. J Am Chem Soc 113(12):4583–4588
10. Voit BI, Lederer A (2009) Hyperbranched and highly branched polymer Architectures synthetic strategies and major characterization aspects. Chem Rev 109(11):5924–5973
11. Korolev GV, Bubnova ML (2007) Synthesis, properties, and practical application of hyperbranched polymers. Polym Sci Ser 49(4):332–354
12. Kim YH, Webster OW (1992) Hyperbranched polyphenylenes. Macromolecules 25 (21):5561–5572
13. Peng H, Dong Y, Jia D, Tang B (2004) Syntheses of readily processable, thermally stable, and light-emitting hyperbranched polyphenylenes. Chin Sci Bull 49(24):2637–2639
14. Tanaka S, Doke Y, Iso T (1997) Preparation of new branched poly(triphenylamine). Chem Commun 21:2063–2064
15. Sun M, Li J, Li B, Fu Y, Bo Z (2005) Toward high molecular weight triphenylamine-based hyperbranched polymers. Macromolecules 38(7):2651–2658
16. Bo Z, Schluter AD (2003) "AB2 + AC2" approach to hyperbranched polymers with a high degree of branching. Chem Commun 18:2354–2355

17. Huang W, Su L, Bo Z (2009) Hyperbranched polymers with a degree of branching of 100% prepared by catalyst transfer suzuki–miyaura polycondensation. J Am Chem Soc 131 (30):10348–10349
18. Maier G, Zech C, Voit B, Komber H (1998) An approach to hyperbranched polymers with a degree of branching of 100%. Macromol Chem Phys 199(12):2655–2664
19. Lim S-J, Seok DY, An BK, Jung SD, Park SY (2006) A modified strategy for the synthesis of hyperbranched poly(p-phenylenevinylene): achieving extended π-conjugation with growing molecular weight. Macromolecules 39(1):9–11
20. Dieck HA, Heck RF (1974) Organophosphinepalladium complexes as catalysts for vinylic hydrogen substitution reactions. J Am Chem Soc 96(4):1133–1136
21. Nishide H, Nambo M, Miyasaka M (2002) Hyperbranched poly(phenylenevinylene) bearing pendant phenoxys for a high-spin alignment. J Mater Chem 12(12):3578–3584
22. Fukuzaki E, Nishide H (2006) Room-temperature high-spin organic single molecule: nanometer-sized and hyperbranched poly[1,2, (4)-phenylenevinyleneanisylaminium]. J Am Chem Soc 128(3):996–1001
23. Lu P, Paulasaari JK, Weber WP (1996) Hyperbranched poly(4-Acetylstyrene) by ruthenium-catalyzed step-growth polymerization of 4-acetylstyrene. Macromolecules 29 (27):8583–8586
24. In I, Lee H, Kim SY (2003) Synthesis of hyperbranched poly(phenylene oxide) by ullmann polycondensation and subsequent utilization as unimolecular micelle. Macromol Chem Phys 204(13):1660–1664
25. Tolosa J, Kub C, Bunz UHF (2009) Hyperbranched: a universal conjugated polymer platform. Angew Chem Int Ed 48(25):4610–4612
26. Li ZA, Wu W, Ye C, Qin J, Li Z (2010) New second-order nonlinear optical polymers derived from AB2 and AB monomers via sonogashira coupling reaction. Macromol Chem Phys 211(8):916–923
27. Fussell AL, Isomaki A, Strachan CJ (2013) Non-linear optical imaging—introduction and pharmaceutical applications. Am Pharmaceut Rev 16(6):54–63
28. Stille JK (1972) Cycloaddition polymerization. Die Makromolekulare Chemie 154(1):49–61
29. Morgenroth F, Mullen K (1997) Dendritic and hyperbranched polyphenylenes via a simple Diels-Alder route. Tetrahedron 53(45):15349–15366
30. Harrison RM, Feast WJ (1997) ACS Polym Mater Sci Eng 77:162
31. Abadie MJM (2012) High performance polymers—polyimides based—from chemistry to applications. InTech:
32. Gok O, Durmaz H, Ozdes ES, Hizal G, Tunca U, Sanyal A (2010) Maleimide-based thiol reactive multiarm star polymers via Diels-Alder/retro Diels-Alder strategy. J Polym Sci A Polym Chem 48(12):2546–2556
33. Sanyal A (2010) Diels-alder cycloaddition-cycloreversion: a powerful combo in materials design. Macromol Chem Phys 211(13):1417–1425
34. Froimowicz P, Frey H, Landfester K (2011) Towards the generation of self-healing materials by means of a reversible photo-induced approach. Macromol Rapid Commun 32(5):468–473
35. Itoya K, Kakimoto M, Imai Y (1994) High-pressure synthesis of new aromatic poly (diazetidinediones) by cyclodimerization polymerization of aromatic diisocyanates. Macromolecules 27(25):7231–7235
36. Xu KT, Tang BZ (1999) Polycyclotrimerization of diynes, a new approach to hyperbranched polyphenylenes. Chin J Polym Sci 17(4):397–402
37. Peng H, Cheng L, Luo J, Xu K, Sun Q, Dong Y, Salhi F, Lee PPS, Chen J, Tang BZ (2002) Simple synthesis, outstanding thermal stability, and tunable light-emitting and optical-limiting properties of functional hyperbranched polyarylenes. Macromolecules 35 (14):5349–5351
38. Jin RH, Motokucho S, Andou Y, Nishikubo T (1998) Controlled polymerization of an AB2 monomer using a chloromethylarene as comonomer: branched polymers from activated methylene compounds. Macromol Rapid Commun 19(1):41–46

39. In I, Kim SY (2005) Hyperbranched poly(arylene ether amide) via nucleophilic aromatic substitution reaction. Macromol Chem Phys 206(18):1862–1869
40. Yang D, Kong J (2016) 100% hyperbranched polymers via the acid-catalyzed friedel-crafts aromatic substitution reaction. Polym. Chem. 7(33):5226–5232
41. Kim Y-B, Kim HK, Nishida H, Endo T (2004) Synthesis and characterization of hyperbranched poly(β-ketoester) by the michael addition. Macromol Mater Eng 289 (10):923–926
42. Trumbo DL (1991) Michael addition polymers from 1,4 and 1,3 benzenedimethanol diacetoacetates and tripropylene glycol diacrylate. Polym Bull 26(3):265–270
43. Gao C, Xu Y, Yan D, Chen W (2003) Water-soluble degradable hyperbranched polyesters: novel candidates for drug delivery? Biomacromol 4(3):704–712
44. Park MR, Han KO, Han IK, Cho MH, Nah JW, Choi YJ, Cho CS (2005) Degradable polyethylenimine-alt-poly(ethylene glycol) copolymers as novel gene carriers. J Control Release 105(3):367–380
45. Cosulich ME, Russo S, Pasquale S, Mariani A (2000) Performance evaluation of hyperbranched aramids as potential supports for protein immobilization. Polymer 41 (13):4951–4956
46. Kim YH (1992) Lyotropic liquid crystalline hyperbranched aromatic polyamides. J Am Chem Soc 114(12):4947–4948
47. Lee CC (2015) The current trends of optics and photonics, vol 129. Springer Netherlands
48. Ishida Y, Sun ACF, Jikei M, Kakimoto M (2000) Synthesis of hyperbranched aromatic polyamides starting from dendrons as ABx monomers: effect of monomer multiplicity on the degree of branching. Macromolecules 33(8):2832–2838
49. Yamanaka K, Jikei M, Kakimoto MA (2000) Synthesis of hyperbranched aromatic polyimides via polyamic acid methyl ester precursor. Macromolecules 33(4):1111–1114
50. Kolb HC, Finn MG, Sharpless KB (2001) Click chemistry: diverse chemical function from a few good reactions. Angew Chem Int Ed 40(11):2004–2021
51. Tornoe CW, Christensen C, Meldal M (2002) Peptidotriazoles on solid phase: [1,2,3]-triazoles by regiospecific copper(I)-catalyzed 1,3-dipolar cycloadditions of terminal alkynes to azides. J Org Chem 67(9):3057–3064
52. Scheel AJ, Komber H, Voit BI (2004) Novel hyperbranched poly([1,2,3]-triazole)s derived from AB2 monomers by a 1,3-dipolar cycloaddition. Macromol Rapid Commun 25 (12):1175–1180
53. Li ZA, Yu G, Hu P, Ye C, Liu Y, Qin J, Li Z (2009) New azo-chromophore-containing hyperbranched polytriazoles derived from ab2 monomers via click chemistry under copper (I) catalysis. Macromolecules 42(5):1589–1596
54. Zhang L, Chen X, Xue P, Sun HHY, Williams ID, Sharpless KB, Fokin VV, Jia G (2005) Ruthenium-catalyzed cycloaddition of alkynes and organic azides. J Am Chem Soc 127 (46):15998–15999
55. van Dijk M, Rijkers DTS, Liskamp RMJ, van Nostrum CF, Hennink WE (2009) Synthesis and applications of biomedical and pharmaceutical polymers via click chemistry methodologies. Bioconjugate Chem 20(11):2001–2016
56. Li H, Wang J, Sun JZ, Hu R, Qin A, Tang BZ (2012) Metal-free click polymerization of propiolates and azides: facile synthesis of functional poly(aroxycarbonyltriazole)s. Polym Chem 3(4):1075–1083
57. Li H, Wu H, Zhao E, Li J, Sun JZ, Qin A, Tang BZ (2013) Hyperbranched poly (aroxycarbonyltriazole)s: metal-free click polymerization, light refraction, aggregation-induced emission, explosive detection, and fluorescent patterning. Macromolecules 46(10):3907–3914
58. Ortega P, Cobaleda BM, Hernandez-Ros JM, Fuentes-Paniagua E, Sanchez-Nieves J, Tarazona MP, Copa-Patino JL, Soliveri J, de la Mata FJ, Gomez R (2011) Hyperbranched polymers versus dendrimers containing a carbosilane framework and terminal ammonium groups as antimicrobial agents. Org Biomol Chem 9(14):5238–5248

59. Xue L, Yang Z, Wang D, Wang Y, Zhang J, Feng S (2013) Synthesis and characterization of silicon-containing hyperbranched polymers via thiol-ene click reaction. J Organomet Chem 732:1–7

60. Moreno S, Lozano-Cruz T, Ortega P, Tarazona MP, de la Mata FJ, Gómez R (2014) Synthesis of new amphiphilic water-stable hyperbranched polycarbosilane polymers. Polym Int 63(7):1311–1323

61. Roy RK, Ramakrishnan S (2011) Thiol-ene clickable hyperscaffolds bearing peripheral allyl groups. J Polym Sci Part APolym Chem 49(8):1735–1744

62. Konkolewicz D, Gray-Weale A, Perrier SB (2009) Hyperbranched polymers by thiol-yne chemistry: from small molecules to functional polymers. J Am Chem Soc 131(50):18075–18077

63. Cook AB, Barbey R, Burns JA, Perrier S (2016) Hyperbranched polymers with high degrees of branching and low dispersity values: pushing the limits of thiol-yne chemistry. Macromolecules 49(4):1296–1304

64. Han J, Zhao B, Tang A, Gao Y, Gao C (2012) Fast and scalable production of hyperbranched polythioether-ynes by a combination of thiol-halogen click-like coupling and thiol-yne click polymerization. Polym Chem 3(7):1918–1925

65. Wang D, Zhao T, Zhu X, Yan D, Wang W (2015) Bioapplications of hyperbranched polymers. Chem Soc Rev 44(12):4023–4071

66. Lasprilla AJR, Martinez GAR, Hoss B (2011) Synthesis and characterization of poly (lactic acid) for use in biomedical field. Chem Eng 24:985–990

67. Tasaka F, Ohya Y, Ouchi T (2001) One-pot synthesis of novel branched polylactide through the copolymerization of lactide with mevalonolactone. Macromol Rapid Commun 22 (11):820–824

68. Pitet LM, Hait SB, Lanyk TJ, Knauss DM (2007) Linear and branched architectures from the polymerization of lactide with glycidol. Macromolecules 40(7):2327–2334

69. Tsujimoto T, Haza Y, Yin Y, Uyama H (2011) Synthesis of branched poly(lactic acid) bearing a castor oil core and its plasticization effect on poly(lactic acid). Polym J 43(4):425–430

70. Frey H (2013) Hyperbranched polyglycerols (Synthesis and Applications). In: Encyclopedia of Polymeric Nanomaterials, Springer Berlin Heidelberg: Berlin, Heidelberg, pp 1–4

71. Wilms D, Stiriba S-E, Frey H (2010) Hyperbranched polyglycerols: from the controlled synthesis of biocompatible polyether polyols to multipurpose applications. Acc Chem Res 43(1):129–141

72. Sunder A, Hanselmann R, Frey H, Mulhaupt R (1999) Controlled synthesis of hyperbranched polyglycerols by ring-opening multibranching polymerization. Macromolecules 32(13):4240–4246

73. Robinson JW, Zhou Y, Bhattacharya P, Erck R, Qu J, Bays JT, Cosimbescu L (2016) Probing the molecular design of hyper-branched aryl polyesters towards lubricant applications. Sci Rep 6:18624

74. Khemchandani B, Verma HS (2012) High performance shear stable viscosity modifiers. In: Polymer processing and characterization, Apple Academic Press, pp 33–41

75. Stohr T, Eisenberg B, Muller M (2008) A new generation of high performance viscosity modifiers based on comb polymers. SAE Int J Fuels Lubr 1(1):1511–1516

76. Lee S, Saito K, Lee HR, Lee MJ, Shibasaki Y, Oishi Y, Kim BS (2012) Hyperbranched double hydrophilic block copolymer micelles of poly(ethylene oxide) and polyglycerol for pH-responsive drug delivery. Biomacromol 13(4):1190–1196

77. Garamus VM, Maksimova TV, Kautz H, Barriau E, Frey H, Schlotterbeck U, Mecking S, Richtering W (2004) Hyperbranched polymers: structure of hyperbranched polyglycerol and amphiphilic poly(glycerol ester)s in dilute aqueous and nonaqueous solution. Macromolecules 37(22):8394–8399

78. Parzuchowski PG, Grabowska M, Jaroch M, Kusznerczuk M (2009) Synthesis and characterization of hyperbranched polyesters from glycerol-based AB2 monomer. J Polym Sci A Polym Chem 47(15):3860–3868

79. Parzuchowski PG, Jaroch M, Tryznowski M, Rokicki G (2008) Synthesis of new glycerol-based hyperbranched polycarbonates. Macromolecules 41(11):3859–3865
80. Testud B, Pintori D, Grau E, Taton D, Cramail H (2017) Hyperbranched polyesters by polycondensation of fatty acid-based ABn-type monomers. Green Chem 19(1):259–269
81. Brenner AR, Voit BI, Massa DJ, Turner SR (1996) Hyperbranched polyesters: End group modification and properties. Macromol Symp 102(1):47–54
82. Ghosh A, Banerjee S, Voit B (2015) Aromatic hyperbranched polymers: synthesis and application. In: Long TE, Voit B, Okay O (eds) Porous carbons- hyperbranched polymers-polymer solvation. Springer International Publishing, Cham, pp 27–124
83. Hult A, Johansson M, Malmstrom E (1999) Hyperbranched Polymers. In: Roovers J (ed) Branched polymers II. Springer Berlin Heidelberg: Berlin, Heidelberg, pp 1–34
84. Zhang X (2010) Hyperbranched aromatic polyesters: From synthesis to applications. Prog Org Coat 69(4):295–309
85. Kricheldorf HR, Zang Q-Z, Schwarz G (1982) New polymer syntheses: 6. Linear and branched poly(3-hydroxy-benzoates). Polymer 23(12):1821–1829
86. Turner SR, Voit BI, Mourey TH (1993) All-aromatic hyperbranched polyesters with phenol and acetate end groups: synthesis and characterization. Macromolecules 26(17):4617–4623
87. Fomine S, Rivera E, Fomina L, Ortiz A, Ogawa T (1998) Polymers from coumarines: 4. Design and synthesis of novel hyperbranched and comb-like coumarin-containing polymers. Polymer 39(15):3551–3558
88. Kricheldorf HR, Stukenbrock T (1998) New polymer syntheses XCIII. Hyperbranched homo- and copolyesters derived from gallic acid and β-(4-hydroxyphenyl)-propionic acid. J Polym Sci APolym Chem 36(13):2347–2357
89. Qiu T, Tang L, Tuo X, Zhang X, Liu D (2001) Study on self-assembly properties of aryl-alkyl hyperbranched polyesters with carboxylic end groups. Polym Bull 47(3):337–342
90. Jikei M, Kakimoto MA (2001) Hyperbranched polymers: a promising new class of materials. Prog. Polym. Sci. 26(8):1233–1285
91. Erber M, Boye S, Hartmann T, Voit BI, Lederer A (2009) A convenient room temperature polycondensation toward hyperbranched AB2-type all-aromatic polyesters with phenol terminal groups. J Polym Sci Polym Chem 47(19):5158–5168
92. Gross RA, Kumar A, Kalra B (2001) Polymer synthesis by in vitro enzyme catalysis. Chem Rev 101(7):2097–2124
93. Uyama H, Kobayashi S (2002) Enzyme-catalyzed polymerization to functional polymers. J Mol Catal B Enzym 19–20:117–127
94. Reihmann M, Ritter H (2006) Synthesis of phenol polymers using peroxidases. In: Kobayashi S, Ritter H, Kaplan D (eds) Enzyme-catalyzed synthesis of polymers. Springer Berlin Heidelberg: Berlin, Heidelberg, pp 1–49
95. Skaria S, Smet M, Frey H (2002) Enzyme-catalyzed synthesis of hyperbranched aliphatic polyesters. Macromol Rapid Commun 23(4):292–296
96. Lopez-Luna A, Gallegos JL, Gimeno M, Vivaldo-Lima E, Barzana E (2010) Lipase-catalyzed syntheses of linear and hyperbranched polyesters using compressed fluids as solvent media. J Mol Catal B Enzym 67(1–2):143–149
97. Mena M, Lopez-Luna A, Shirai K, Tecante A, Gimeno M, Barzana E (2013) Lipase-catalyzed synthesis of hyperbranched poly-l-lactide in an ionic liquid. Bioproc Biosyst Eng 36(3):383–387
98. Xu F, Zhong J, Qian X, Li Y, Lin X, Wu Q (2013) Multifunctional poly(amine-ester)-type hyperbranched polymers: lipase-catalyzed green synthesis, characterization, biocompatibility, drug loading and anticancer activity. Polym Chem 4(12):3480–3490
99. Kricheldorf H (2013) Hyperbranched polymers by a2 + bn polycondensation. In: Polycondensation: history and new results. Springer Berlin Heidelberg: Berlin, Heidelberg, pp 147–159
100. Aharoni SM, Edwards SF (1989) Gels of rigid polyamide networks. Macromolecules 22 (8):3361–3374

101. Aharoni SM (1991) Gels of two-step rigid polyamide networks. Macromolecules 24 (15):4286–4294

102. Jikei M, Chon SH, Kakimoto MA, Kawauchi S, Imase T, Watanebe J (1999) Synthesis of hyperbranched aromatic polyamide from aromatic diamines and trimesic acid. Macromolecules 32(6):2061–2064

103. Fang J, Kita H, Okamoto KI (2000) Hyperbranched polyimides for gas separation applications. 1. Synthesis and characterization. Macromolecules 33(13):4639–4646

104. Hao J, Jikei M, Kakimoto MA (2002) Preparation of hyperbranched aromatic polyimides via A2 + B3 Approach. Macromolecules 35(14):5372–5381

105. Unal S, Long TE (2006) Highly Branched Poly(ether ester)s via cyclization-free melt condensation of A2 oligomers and B3 monomers. Macromolecules 39(8):2788–2793

106. Scheel A, Komber H, Voit B (2004) Hyperbranched thermolabile polycarbonates derived from a A2 + B3 monomer system. Macromol Symp 210(1):101–110

107. Miyasaka M, Takazoe T, Kudo H, Nishikubo T (2010) Synthesis of hyperbranched polycarbonate by novel polymerization of di-tert-butyl tricarbonate with 1,1,1-tris (4-hydroxyphenyl)ethane. Polym J 42(11):852–859

108. Wang Q, Shi W (2006) Synthesis and thermal decomposition of a novel hyperbranched polyphosphate ester used for flame retardant systems. Polym Degrad Stab 91(6):1289–1294

109. Liu J, Huang W, Pang Y, Yan D (2015) Hyperbranched polyphosphates: synthesis, functionalization and biomedical applications. Chem Soc Rev 44(12):3942–3953

110. Xie J, Hu L, Shi W, Deng X, Cao Z, Shen Q (2008) Synthesis and characterization of hyperbranched polytriazole via an 'A2 + B3' approach based on click chemistry. Polym Int 57(8):965–974

111. Qin A, Lam JWY, Jim CKW, Zhang L, Yan J, Haussler M, Liu J, Dong Y, Liang D, Chen E, Jia G, Tang BZ (2008) Hyperbranched Polytriazoles: click polymerization, regioisomeric structure, light emission, and fluorescent patterning. Macromolecules 41(11):3808–3822

112. Chen H, Jia J, Duan X, Yang Z, Kong J (2015) Reduction-cleavable hyperbranched polymers with limited intramolecular cyclization via click chemistry. J Polym Sci A Polym Chem 53(20):2374–2380

113. Tapan K, Thirumoolan D, Mohanram R, Vetrivel K, Basha KA (2015) Antimicrobial activity of hyperbranched polymers: synthesis, characterization, and activity assay study. J Bioact Compat Polym 30(2):145–156

114. Gao C, Yan D (2004) Hyperbranched polymers: from synthesis to applications. Prog Polym Sci 29(3):183–275

115. Gao C, Yan D (2011) Synthesis of Hyperbranched polymers via polymerization of asymmetric monomer pairs. In: Hyperbranched Polymers, John Wiley & Sons, Inc. pp 107–138

116. Shi Y, Graff RW, Gao H (2015) Recent progress on synthesis of hyperbranched polymers with controlled molecular weight distribution. In: Controlled Radical Polymerization: Materials, American Chemical Society, vol. 1188, pp 135–147

117. Feast WJ, Stainton NM (1995) Synthesis, structure and properties of some hyperbranched polyesters. J Mater Chem 5(3):405–411

118. Bharathi P, Moore JS (2000) Controlled synthesis of hyperbranched polymers by slow monomer addition to a core. Macromolecules 33(9):3212–3218

119. Yan D, Zhou Z (1999) Molecular weight distribution of hyperbranched polymers generated from polycondensation of AB2 type monomers in the presence of multifunctional core moieties. Macromolecules 32(3):819–824

120. Radke W, Litvinenko G, Muller AHE (1998) Effect of core-forming molecules on molecular weight distribution and degree of branching in the synthesis of hyperbranched polymers. Macromolecules 31(2):239–248

121. Satoh T (2012) Synthesis of hyperbranched polymer using slow monomer addition method. Int J Polym Sci p 8

122. Chen JY, Smet M, Zhang JC, Shao WK, Li X, Zhang K, Fu Y, Jiao YH, Sun T, Dehaen W, Liu FC, Han EH (2014) Fully branched hyperbranched polymers with a focal point: analogous to dendrimers. Polym Chem 5(7):2401–2410
123. Schull C, Frey H (2013) Grafting of hyperbranched polymers: from unusual complex polymer topologies to multivalent surface functionalization. Polymer 54(21):5443–5455
124. Schull C, Rabbel H, Schmid F, Frey H (2013) Polydispersity and molecular weight distribution of hyperbranched graft copolymers via "hypergrafting" of abm monomers from polydisperse macroinitiator cores: theory meets synthesis. Macromolecules 46(15):5823–5830
125. Popeney CS, Lukowiak MC, Bottcher C, Schade B, Welker P, Mangoldt D, Gunkel G, Guan Z, Haag R (2012) Tandem coordination, ring-opening, hyperbranched polymerization for the synthesis of water-soluble core-shell unimolecular transporters. ACS Macro Lett 1(5):564–567
126. Bharathi P, Moore JS (1997) Solid-Supported Hyperbranched Polymerization: Evidence for Self-Limited Growth. J Am Chem Soc 119(14):3391–3392
127. Zhou Z, Jia Z, Yan D (2012) Kinetic analysis of AB2 polycondensation in the presence of multifunctional cores with various reactivities. Polymer 53(15):3386–3391
128. Bernal DP, Bedrossian L, Collins K, Fossum E (2003) Effect of core reactivity on the molecular weight, polydispersity, and degree of branching of hyperbranched poly(arylene ether phosphine oxide)s. Macromolecules 36(2):333–338
129. Roy RK, Ramakrishnan S (2011) Control of molecular weight and polydispersity of hyperbranched polymers using a reactive B3 core: a single-step route to orthogonally functionalizable hyperbranched polymers. Macromolecules 44(21):8398–8406
130. Suzuki M, Ii A, Saegusa T (1992) Multibranching polymerization: palladium-catalyzed ring-opening polymerization of cyclic carbamate to produce hyperbranched dendritic polyamine. Macromolecules 25(25):7071–7072
131. Suzuki M, Yoshida S, Shiraga K, Saegusa T (1998) New ring-opening polymerization via a π-allylpalladium complex. 5. multibranching polymerization of cyclic carbamate to produce hyperbranched dendritic polyamine. Macromolecules 31(6):1716–1719
132. Ohta Y, Fujii S, Yokoyama A, Furuyama T, Uchiyama M, Yokozawa T (2009) Synthesis of well-defined hyperbranched polyamides by condensation polymerization of AB2 monomer through changed substituent effects. Angew Chem Int Ed 48(32):5942–5945

Chapter 3
Part II—Synthesis of Hyperbranched Polymers: Mixed Chain-Growth and Step-Growth Methods

Abbreviations

ACDT	2-((2-(((dodecylthiocarbonothioyl)thio)-2-methylpropanoyl)oxy)-ethyl acrylate
AIBN	Azobisisobutyronitrile
AIG1c	3-O-acryloyl-1,2,5,6-di-O-isopropylidene-α-D-glucofuranoside
ARGET	Activators regenerated by electron transfer
ATRA	Atom transfer radical addition
ATRP	Atom transfer radical polymerization
BPEA	2-((2-bromopropionyl)oxy)ethyl acrylate
BuLi	Butyl lithium
CT	Chain transfer
CTA	Chain transfer agent
DB	Degree of branching
DMAEMA	2-(dimethylamino)ethyl methacrylate
DVA	Divinyl adipate
DVB	Divinyl benzene
DVM	Divinyl monomer
ECTVA	Vinyl 2-(ethoxycarbonothioylthio) acetate
EGDMA	Ethylene glycol dimethacrylate
EHMO	3-ethyl-3-hydromethyl oxetane
FRP	Free radical polymerization
GTP	Group transfer polymerization
ICAR	Initiators for continuous activator regeneration
IFIRP	Initiator fragment incorporation radical polymerization
MMA	Methyl methacrylate
MVM	Multivinyl polymerization
M.W	Molecular weight
M.W.D	Molecular weight distribution
NMP/NMRP	Nitroxide mediated radical polymerization
PB	Poly (buta-1,2-diene)
PAMAM	Poly (amido amine)
PCS	Poly (carbosilane)

© Springer Nature Singapore Pte Ltd. 2018
A. Bandyopadhyay et al., *Hyperbranched Polymers for Biomedical Applications*,
Springer Series on Polymer and Composite Materials,
https://doi.org/10.1007/978-981-10-6514-9_3

P.D.I	Poly dispersity index
PE	Polyethylene
PEG	Polyethylene glycol
PEGDMA	Poly ethylene glycol dimethacrylate
PEI	Poly (ethylene imine)
PEO	Polyethylene oxide
PG	Polyglycerol
PMA	Poly (mecthacrylate)
PS	Polystyrene
PVA	Poly (vinyl alcohol)
PVAc	Poly (vinyl acetate)
PRE	Persistant radical effect
PTP	Proton transfer polymerization
RAFT	Reversible addition fragmentation chain transfer
RBC	Red blood cell
ROMBP	Ring opening multi-branching polymerization
SARA-ATRP	Supplementary activators and reducing agents ATRP
SCGTCP	Self-condensing group transfer copolymerization
SCROP	Self-condensing ring opening polymerization
SCVP	Self-condensing vinyl polymerization
SCVCP	Self-condensing vinyl copolymerization
tBMA	Tert-butyl methacrylate
TEMPO	2,2,6,6-tetramethylpiperidin-1-oxyl
TIPNO	2,2,5-trimethyl-4-phenyl-3-azahexane-3-nitroxide
VBC	Vinylbenzyl chloride
VBP	Vasculature binding peptide
WBC	White blood cell or leukocyte

3.1 Introduction to Simultaneous Step- and Chain-Growth Methodologies

With the growing interest and demand in the realm of hyperbranched (hb) polymers, lot of synthesis approaches have already been explored, some of which are detailed earlier in the Chap. 2. Both step-growth and chain-growth approaches are widely followed in the synthesis of hb polymers. Step-growth approaches mainly include AB_x polycondensation and double monomer (symmetric/asymmetric pairs) methodologies. Whereas chain-growth approaches include radical copolymerization, surface grafting, and other controlled polymerization techniques. Interestingly, both self-condensing vinyl polymerization (SCVP) and self-condensing ring opening polymerization follow step-growth as well as chain-growth routes. In step-growth approaches, high DP is achieved only at high monomer conversions (as predicted from Carother's equation). Hence, often hb polymers with high M.W cannot be obtained as high monomer conversions are avoided in practical cases, which otherwise would encourage gelation. On the contrary, in chain-growth approaches as the polymer chains grow via reactions of the growing chain ends with monomers, polymerization reaction completes in a reasonable time (unlike step growth) and thus ensures high DP/high M.W. However, chain-growth polymerization is highly random in nature which eventually broadens M.W.D of the resultant hb polymers. Hence, it would be highly desirable to perform chain-growth and step-growth polymerization simultaneously in order to ensure hb polymers with controlled properties (especially desirable for biomedical applications). This chapter deals with more recent and controlled synthesis strategies based on both chain-growth and step-growth approaches that are used in the development of hb polymers for biomedical applications.

3.2 Radical Polymerization

In the preparation of hb polymers via free radical polymerization (FRP), chain transfer to monomer, polymer, or other molecules is a guiding parameter and must be significantly high. Transfer to polymer is significant in ethylene and vinyl acetate systems, whereas transfer to monomer is popular with the acrylate systems.

Liu et al. reported synthesis of a hb polystyrene via FRP of styrene and a comonomer 4-vinyl benzyl thiol (which is a chain transfer, CT monomer) [1]. A CT monomer is constituted of a vinyl group and a telogen group (generally a thiol group). During a FRP, low M.W oligomers containing terminal double bonds are generated which subsequently act as macromonomers and allow further polymerization to limited chain lengths (due to telomerization of the telogen groups); Scheme 3.1. This process of the generation of macromonomers and subsequent FRP through telomerization is analogous to AB$_x$ polycondensation (which even follows Flory's branching theory) and thus generates hb polymers [1]. In spite of the high potentiality of CT monomers in developing hb polymers, they are often discouraged. CT monomers are unstable and susceptible to self-condensation by Michael addition. And also the availability of CT monomer is highly limited. In this regard, usage of a balancing amount of a chain transfer agent (CTA); usually a long chained thiol, in the recipe of a vinyl polymerization reaction, facilitates transfer reactions and in turn encourages formation of the hb polymers from a myriad of vinyl monomers in the presence of a multivinyl monomer (MVM); Scheme 3.2. This concept gave birth to a new synthesis strategy, named as "Strathclyde methodology" which was introduced in the year 2000, by the gang of Sherrington from the University of Strathclyde, UK [2]. "Strathclyde methodology" is easy to follow and cost-effective which encourages commercialization of the hb polymers. Hb polymers from different vinyl monomers have already been prepared by following the "Strathclyde methodology" [3–5]. In fact, in our lab, we have successfully developed amphiphilic hb copolymers of acrylic acid and propargyl

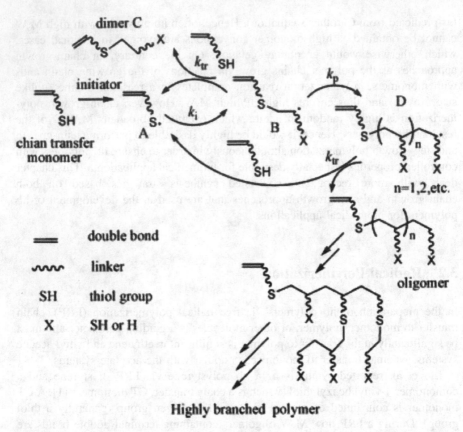

Highly branched polymer

Scheme 3.1 Schematic representation of the generation of a hb polymer from a CT monomer. Reprinted (adapted) with permission from Liu et al. [1]. Copyright (2008) Wiley Online Library

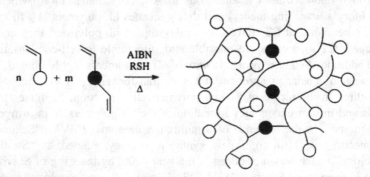

Scheme 3.2 Schematic representation of "Strathclyde methodology" for the synthesis of a hb polymer. Reprinted (adapted) with permission from O'Brien et al. [2]. Copyright (2000) Elsevier

acrylate through the "Strathclyde methodology" and subsequently functionalized them through click reaction with a surfactant [6]. The resultant hb copolymer was capable of self-assembling from polymersomes to aggregates when pH of the medium was decreased from an alkaline phase to an acidic phase which facilitated encapsulation of both hydrophilic and hydrophobic molecules. Typically, in the recipe of "Strathclyde methodology", MVM is basically a crosslinking agent. Incorporation of MVM into a FRP recipe produces crosslinked networks (as the number of crosslinks per primary chain exceeds unity even at a low concentration of MVM and at low monomer conversions). Depending upon the level of monomer dilution, FRP of a vinyl monomer in the presence of MVM either produces macrogelation (in concentrated solutions) or else causes microgelation (in dilute solutions). Hence, branched polymers are thought to be precursors to crosslinked polymers and the former can be isolated only at an infinite monomer dilution (which is a hypothetical case). One of the ways of preventing gelation and ensuring hb structures is by keeping the monomer conversion incomplete, so that the vinyl groups along the primary polymer chains are not fully consumed; some vinyl groups remain free. However, in a FRP even if controlled, it is impossible to avoid intramolecular cyclization which causes consumption of the free vinyl groups along the primary polymer chains. Hence, pure branched polymers could never be obtained through "Strathclyde methodology" but are popular only due to cost-effectiveness of the process. In this regard, some research groups employed asymmetrical MVMs for the synthesis of hb polymers with pendant vinyl groups [7, 8]. However, again, this approach lacks versatility owing to the limitations in the design and availability of the asymmetrical MVMs. In a MVM system, a critical overlap polymer chain concentration c* determines the nature of the dominant interactions [9]. When the concentration is below c* then intramolecular cyclization is favored whereas above c* intermolecular branching is favored. Raising the primary polymer chain concentration (say by reducing solvent proportion) increases the risk of gelation. In this respect, "Strathclyde methodology" is highly useful as it helps in maintaining a desirable primary polymer chain length to ensure hb structures through controlled chain transfer reactions by thiols. Typically, in a recipe for "Strathclyde methodology", concentration of MVM should be maintained below 15% and molar ratio of MVM to initiator should not exceed 1 in order to prevent gelation. Unfortunately, due to randomness of the "Strathclyde methodology" (nonliving nature) which adversely affects the properties of the hb polymers, due to usage of toxic thiols/MVMs and also due to lack of variety in desirable functionalities, this approach is hardly followed to synthesize hb polymers for biomedical applications. Due to the nonliving nature of the "Strathclyde methodology", i.e., occurrence of irreversible chain transfer reactions, near-diffusion controlled radical coupling and disproportionation, hb polymers with broad M.W.D, and uncontrolled/low DB are produced. However, recently, attractive post-polymer functionalization (say via click reaction) is attracting further research in the development of "Strathclyde methodology" in spite of the shortcomings of the process.

In many instances, apart from using thiols, transition metal catalysts (generally low spin cobalt (II) complexes like cobaloxime) have also been used to control chain transfer reactions in FRP for the generation of hb polymers [10]. In fact, catalytic chain transfer polymerization (CCTP) is an extension of "Strathclyde methodology". CCTP or rather "cascade polymerization" forms macromonomers, i.e., polymers carrying vinyl ω-end groups, which eventually facilitate post-polymerization end group modification. In general, CCTP follows a two-step radical process; Scheme 3.3 [11]. In the first step, Co(II) complex abstracts a ß-hydrogen atom from the propagating radical which forms a Co(III)-H complex and a dead polymer chain containing a vinyl ω-end group. The hydrogen abstraction by Co(II) complexes occurs via Co···H···C transition state. The ß-hydrogen abstraction occurs either at an α-methyl position (in case of methacrylates, α-methyl styrene, methacrylonitile, etc.) or at the backbone (for acrylates, styrenics, acrylonitrile, etc.). It has been found that, hydrogen abstraction at the α-methyl position occurs readily and thus monomers containing α-methyl groups are highly active for the CCT reactions. Finally, in the next step, Co(III)-H complex reacts with the monomer to form back an active Co(II) complex and a monomer radical which is capable of propagation. In this type of polymerization, branching topology is maintained through the competition between propagation and chain transfer reactions. Guan in his novel work used "cascade polymerization" of ethylene glycol dimethacrylate (EGDMA, **1**), in the presence of cobalt oxime boron fluoride (COBF) catalyst, to develop hb PEGDMA, **2**; Scheme 3.4 [10]. The homopolymerization of EGDMA resulted in repetitive trimerization of the dimethacrylates which eventually formed the hb polymers. In another work, Smeets et al. prepared hb polymers via Co(II) mediated emulsion copolymerization of MMA and EGDMA [12]. They studied how the physiochemical properties of Co(II) complex, the intrinsic chain transfer activity, and the partitioning behavior played an important role in the development of desirable polymer architecture (branched or crosslinked). There are few advantages of CCTP over the "Strathclyde methodology" which include (1) better control in the polymerization conditions, thereby yielding hb polymers with narrow P.D.I/high DB and (2) presence of significant number of pendant vinyl groups which can be functionalized to desirable groups in the further steps. In case of the "Strathclyde methodology", pendant vinyl groups often undergo organic reactions with thiols and thus their availability decreases [12].

Another variation of the "Strathclyde methodology" for the synthesis of hb polymers is initiator fragment incorporation radical polymerization (IFIRP) which

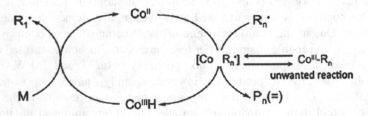

Scheme 3.3 Schematic representation of a catalytic cycle for Co(II) mediated CCT reaction. Reprinted (adapted) with permission from Smeets [11]. Copyright (2013) Elsevier

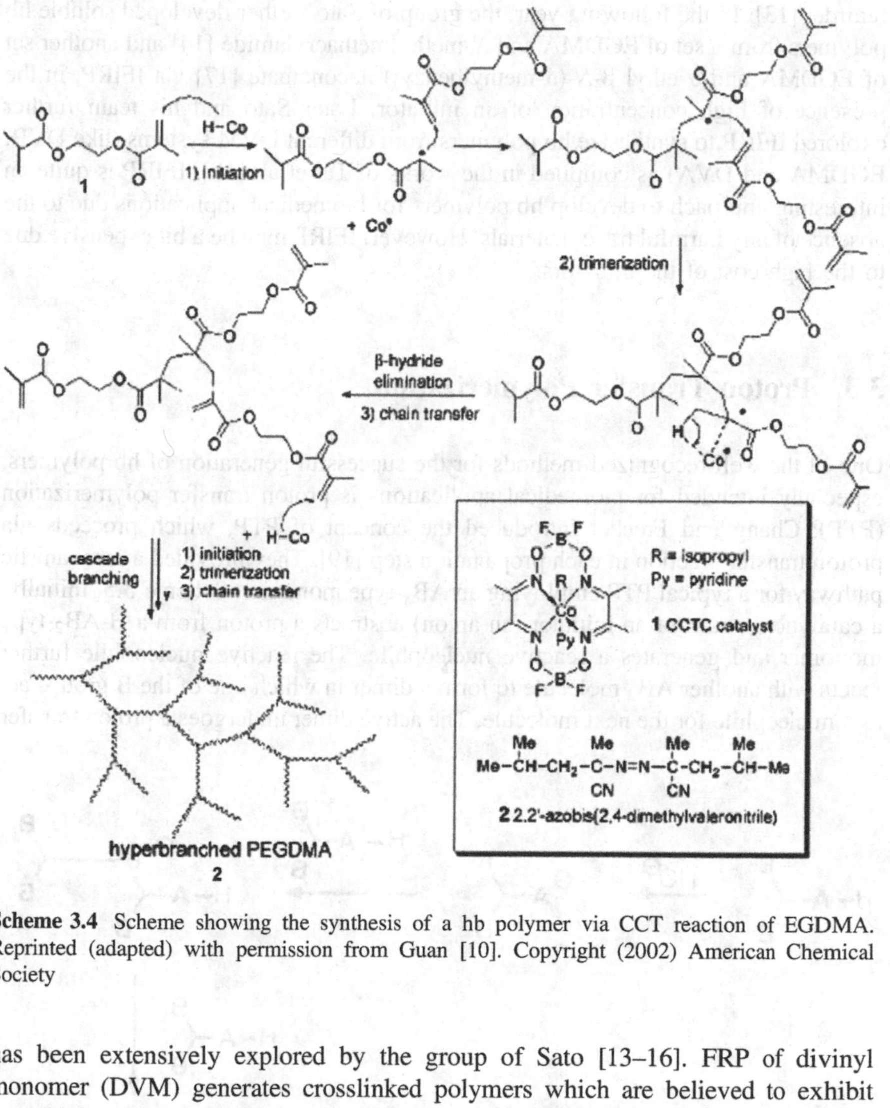

Scheme 3.4 Scheme showing the synthesis of a hb polymer via CCT reaction of EGDMA. Reprinted (adapted) with permission from Guan [10]. Copyright (2002) American Chemical Society

has been extensively explored by the group of Sato [13–16]. FRP of divinyl monomer (DVM) generates crosslinked polymers which are believed to exhibit very high or infinite M.Ws. In a conventional FRP, as the concentration of the initiator increases, M.W of the resultant polymer decreases due to faster chain termination. This fact led to the belief that if the initiator concentration in a FRP of DVM system is kept very high such that M.W decreases dramatically which eventually produces soluble hb polymers. This synthesis approach is named as initiator fragment IFIRP because around 30–40 mol% of the initiator fragments get incorporated into the primary polymer chains through initiation and primary radical termination reactions. In one of the early works, Sato et al. developed organic solvent soluble hb copolymers from concentrated divinyl benzene, DVB and ethylstyrene (EtSt), in the presence of high concentrations of an initiator and a retarder [13]. In the following year, the group of Sato further developed soluble hb polymers from a set of EGDMA and *N*-methylmethacrylamide [14] and another set of EGDMA and α-ethyl β-*N*-(ά-methylbenzyl) itaconamate [17] via IFIRP, in the presence of high concentration of an initiator. Later Sato and his team further explored IFIRP to synthesize hb polymers from different DVM systems (like DVB, EGDMA and DVA) as compiled in the works of Tai et al. [18]. IFIRP is quite an interesting approach to develop hb polymers for biomedical applications due to the absence of any harmful trace materials. However, IFIRP may be a bit expensive due to the high cost of the initiators.

3.3 Proton Transfer Polymerization

One of the well-recognized methods for the successful generation of hb polymers, especially intended for biomedical applications is proton transfer polymerization (PTP). Chang and Frechet introduced the concept of PTP, which proceeds via proton transfer reaction in each propagation step [19]. They provided a mechanistic pathway for a typical PTP employing an AB_x-type monomer; Scheme 3.5. Initially, a catalytic amount of an initiator (an anion) abstracts a proton from a H-AB_2-type monomer and generates a reactive nucleophile. The reactive nucleophile further reacts with another AB_2 molecule to form a dimer in which one of the B groups act as a nucleophile for the next molecule. The active dimer undergoes a proton transfer

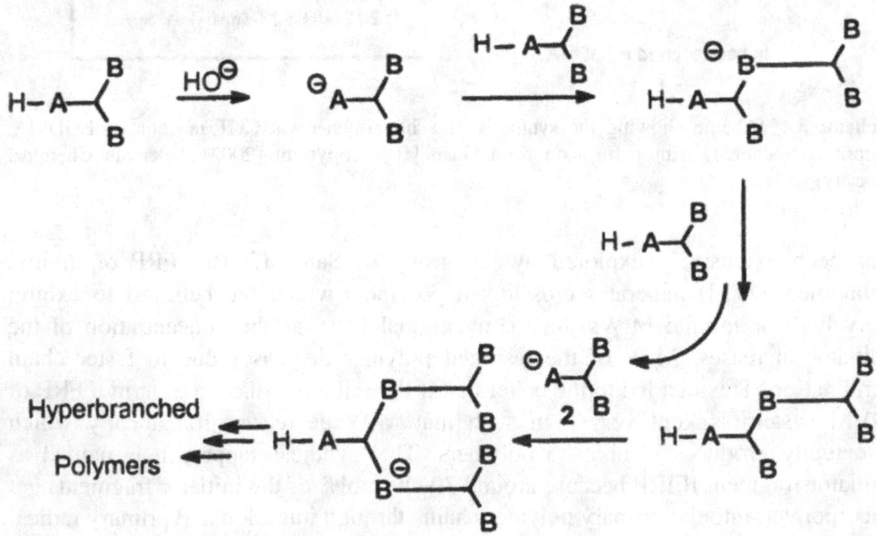

Scheme 3.5 Scheme showing the step-by-step mechanism for the synthesis of a hb polymer via PTP. Reprinted (adapted) with permission from Chang et al. [19]. Copyright (1999) American Chemical Society

reaction with another H-AB$_2$ monomer instead of a propagation reaction because the former is thermodynamically more favorable. Finally, a polymer is formed through subsequent nucleophilic addition and proton transfer reactions. Owing to the multiplicity of the reactive B species in each of the growing molecule, branching is ensured during polymerization. PTP is successful only when the activation of AB$_2$-type monomer or H-A groups of the growing species is significantly faster than the propagation of nucleophile. PTP can be followed using either AB$_x$-type monomers or A$_2$+B$_3$ monomer systems. Generally, PTP is popular with addition and ring opening reactions.

The first notable application of PTP was followed in the synthesis of hb poly (hydroxyether)s from an AB$_2$-type monomer constituting of two epoxide moieties and one phenolic hydroxyl group [19]. The polymerization proceeded via a step-growth route where deprotonation of the phenol was succeeded by nucleophilic attack onto the epoxide ring. Ring opening of the epoxide groups generated an active dimer constituting of a secondary alkoxide anion. In the subsequent step, the activated dimer underwent a proton transfer reaction with another AB$_2$ molecule to regenerate a phenolate anion and a neutral anion. In this reaction, Chang and Frechet maintained the pK$_a$ value of the phenolic group around 10 whereas that of the alkoxide anion was around 17, so that a rapid proton transfer from the phenolic group to the secondary alkoxide group was exclusively possible. The resultant hb poly (hydroxyether)s exhibited a M.W around 2.06×10^5 g mol^{-1}. In another work, Gong and Frechet developed epoxy terminated hb polyesters, **4** from an AB$_2$-type monomer, **3** via PTP; Scheme 3.6 [20].

Chen et al. synthesized a temperature-responsive hb polyether via PTP between 1,2,7,8-diepoxyoctane (DEO) and multiols like ethylene glycol, diethylene glycol, triethylene glycol, 1,2-propane diol and glycerol [21]. The LCSTs of these polymers were tuned in the range of 23.6–67.2 °C just by adjusting the hydrophilic/hydrophobic balance of DEO and multiols, the feed ratio of DEO to multiols and M.Ws. It is worthy to mention at this point that the temperature-responsive polymers are highly useful in drug delivery, separation processes, and sensing applications [22].

In a recent work, Gadwal et al. developed a polythioether-based hb polymer, **6** via base catalyzed PTP of an AB$_2$-type monomer, **5** constituting of a thiol and two epoxides groups; Scheme 3.7 [23]. The hb polymer **6** exhibited DB around 0.65–0.69. The hb polymer **6** contained two reactive sites; hydroxyl units and epoxide units which were distributed throughout the branched structure. The epoxide rings were capable of attaching a lipophilic alkyl, aryl (known for excellent membrane penetration properties), or ethylene oxide group (for excellent stealth properties) through thiol-epoxy reactions while the hydroxyl groups could be attached to the positively charged primary ammonium groups (known for efficient complexation with siRNA molecules). Post-polymer functionalizations of the dual reactive hb polymer **6** yielded dual functionalized hb polymers with high potentiality in gene delivery applications.

As already mentioned that PTP is explored in A$_2$+B$_3$ synthesis strategies, Emrick et al. developed a hb aliphatic polyether, **8** via PTP of 1,2,7,8-

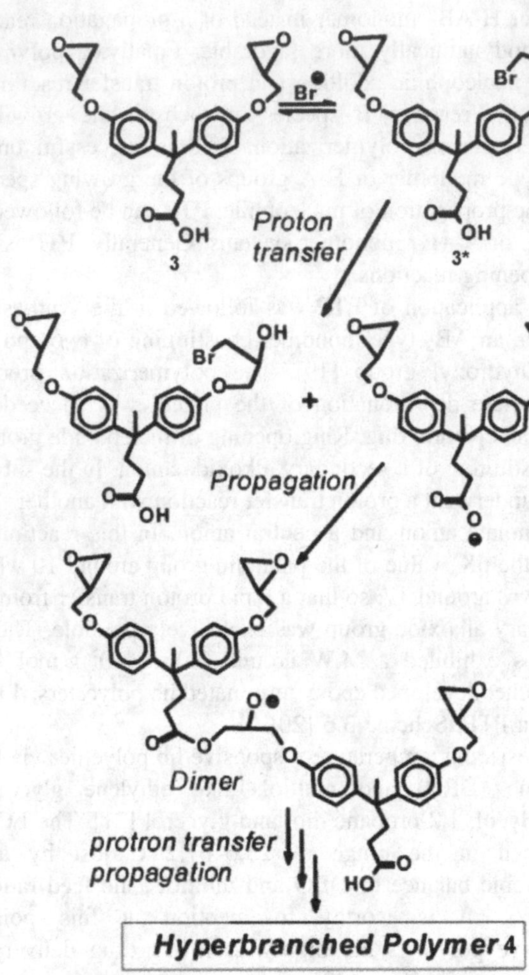

Scheme 3.6 Scheme showing the synthesis of a hb polyester via PTP of an AB₂-type monomer. Reprinted (adapted) with permission from Gong and Frechet [20]. Copyright (2000) American Chemical Society

diepoxyoctane (A₂) and 1,1,1- tris (hydroxymethyl)ethane (B₃) at 120 °C, in the presence of tetra-n-butylammonium chloride as a nucleophilic catalyst; Scheme 3.8 [24]. The hb polymer **8** exhibited a M.W of 15,000 kDa. In another work, Ma et al. synthesized epoxy terminated hb polymers via A₂+B₃ based PTP, from bisphenol-A (A₂) and trimethylolpropane triglycidyl ether (B₃) [25]. From the reported studies, it is clear that PTP is quite a welcome approach in the generation of hb polymers as most of them proceed via addition or ring opening reactions for which numerous peripheral functional groups could be included into the structures (for modification to attract biomedical applications).

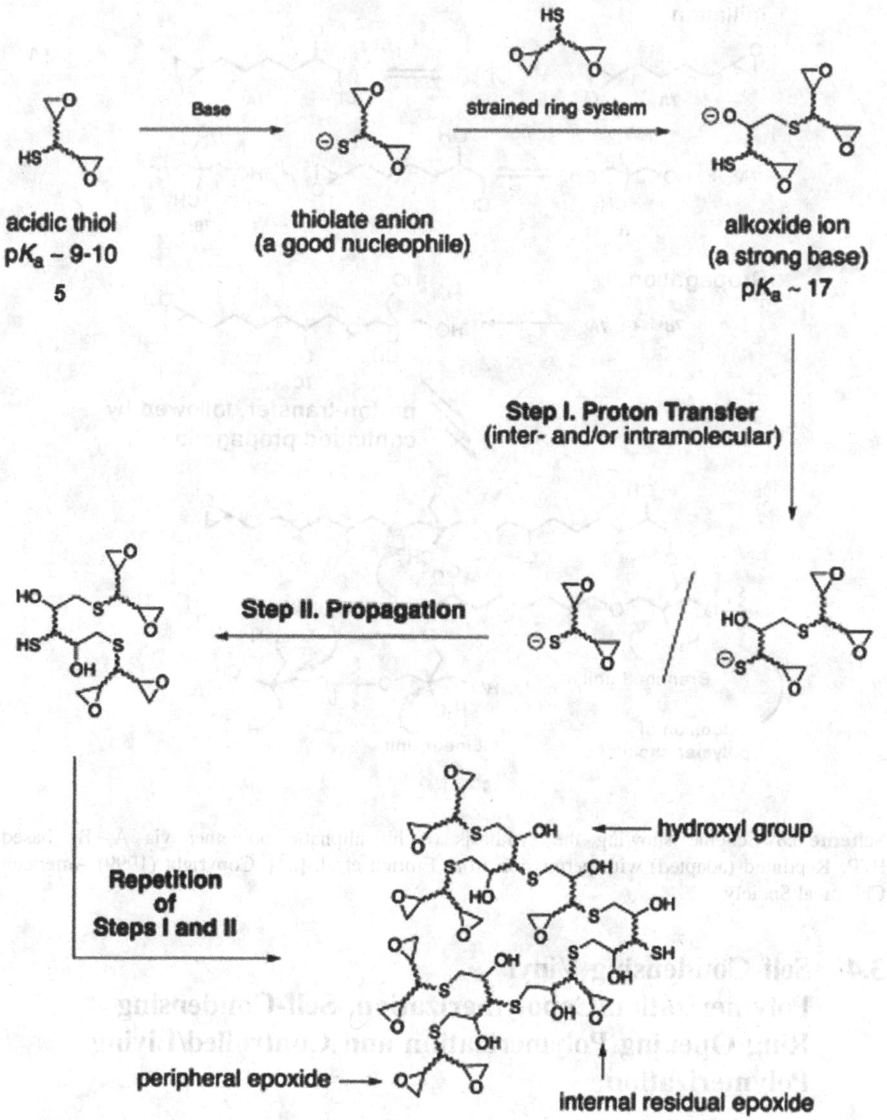

HS
Base
strained ring system
acidic thiol
pK_a ~ 9-10
5
thiolate anion
(a good nucleophile)
alkoxide ion
(a strong base)
pK_a ~ 17
Step I. Proton Transfer
(inter- and/or intramolecular)
Step II. Propagation
Repetition
of
Steps I and II
hydroxyl group
peripheral epoxide
internal residual epoxide
6
Dual-Reactive Hyperbranched Scaffold

Scheme 3.7 Scheme showing the synthesis of a polythioether-based hb polymer via PTP of thiol and epoxide groups. Reprinted (adapted) with permission from Gadwal et al. [23]. Copyright (2014) American Chemical Society

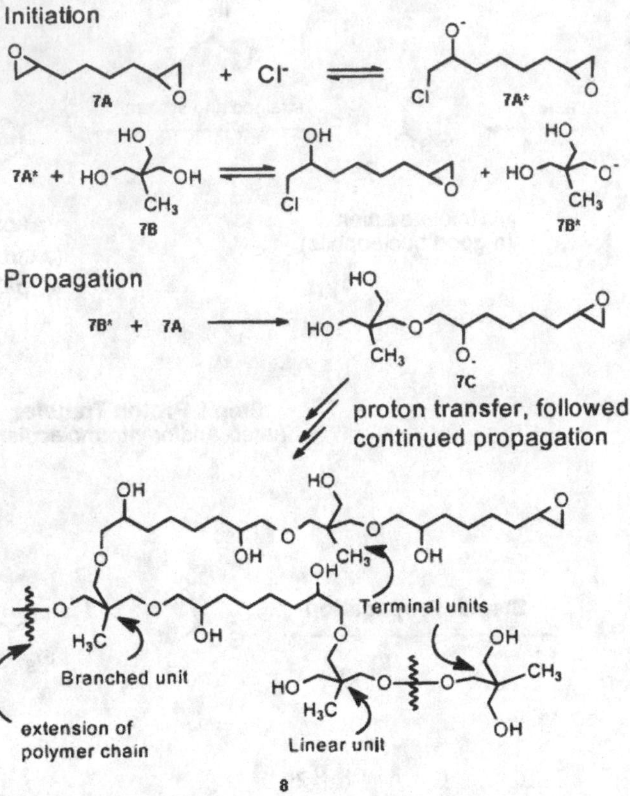

Scheme 3.8 Scheme showing the synthesis of hb aliphatic polyether via A_2+B_3 based PTP. Reprinted (adapted) with permission from Emrick et al. [24]. Copyright (1999) American Chemical Society

3.4 Self-Condensing Vinyl Polymerization/Copolymerization, Self-Condensing Ring Opening Polymerization and Controlled/Living Polymerization

The polymerization of vinyl monomers is still considered as one of the important organic reactions for mass scale productions. However, vinyl monomers cannot be polymerized via AB_x polycondensation. Conventional radical polymerization may produce hb polymers (if chain length is maintained below a critical point) but they exhibit uncontrolled topologies and properties. For biological and biomedical applications like drug delivery, tissue engineering, bioimaging, etc., functional polymers with controlled architectures and properties are essential. One such important functional polymer from nature is protein. Surprisingly, protein is an example of highly stereospecific polymer of amino acids, which exhibits controlled polymer compositions, functionalities, chemical properties, M.W, and uniform

M.W.D. Inspired from nature, attempts have been made to develop polymers with as much control in the macromolecular engineering as possible. Hence, in order to generate hb polymers (with controlled structures) from vinyl monomers, a new category of reaction named as SCVP was introduced under the guidance of Frechet [27]. SCVP employs AB-type initiator monomers (later called inimers). These monomers possess both initiation and propagation properties. In a typical AB-type monomer, A is a vinyl group and B is a pendant group that can be converted into an initiating center B* by the activation of B group in the presence of an external stimulus (Lewis acid or light). The generated B* is capable of reacting with the vinyl groups of another AB molecule to facilitate the propagation reaction. The first reaction of AB* with a double bond group of another AB molecule generates a dimer which acts an AB_2-type intermediate due to the presence of two active initiating sites (A*, B*) and one double bond. Both B* (initiating site) and A* (propagating site) can further react in a repetitive fashion to form three-dimensional polymers, rich in end groups and active centers. Hence, SCVP follows two modes of polymerization—polymerization of double bonds (chain growth) and condensation of the initiating group with double bonds (step growth); Scheme 3.9. Earlier SCVP was believed to follow Flory's branching theory of equal reactivity as that is followed in AB_x polycondensation. However, in recent studies it has been found that in SCVP, the kinetic activities of the chain propagation of the growth sites and the initiating sites differ significantly, which results in a lower DB than those obtained from an AB_x polycondensation reaction. Till date, the theoretical maximum DB of a SCVP reaction is 46.5% [28]. In SCVP, often side reactions like chain transfer or recombination reactions cause crosslinking/or gelation and broaden M.W.D of the resultant hb polymers. Thus, as a remedy, living/controlled polymerization techniques like cationic, anionic, group transfer polymerization (GTP), nitroxide mediated radical polymerization (NMRP), atom transfer radical polymerization (ATRP), reversible addition fragmentation chain transfer polymerization (RAFT), and photo-initiated radical polymerization techniques are preferred. The first SCVP approach as reported by Frechet et al. was based on cationic polymerization where they used 3-(1-chloroethyl)-ethenylbenzene as an AB-type monomer [27]. They employed $SnCl_4$ (as a Lewis acid) in CH_2Cl_2 at −15 or −20 ° C, in the presence of tetrabutylammonium bromide to carry out the cationic polymerization. In this study, they found that M.W-time profile resembled that of the polycondensation process; a slow initial increase in M.W was followed by an exponential growth with time. The resultant hb polymer exhibited a M.W of 2.5×10^5 g mol^{-1} and a P.D.I of 6. Later, SCVP was developed using living cationic copolymerization. Paulo and Puskas synthesized hb polyisobutylenes via living cationic SCVCP of 4-(2-methoxyisopropyl)styrene(p-methoxycumylstyrene) as an AB-type monomer and isobutylene, in the presence of $TiCl_4$ (a co-initiator) in 60:40 mixture of methylcyclohexane and methyl chloride at −80 °C [29]. The obtained hb polymer had a M.W of 8×10^5 g mol^{-1} and a M.W.D of 1.2.

Other variations like SCVP-based anionic polymerization or rather living anionic SCVCP are also followed. Knauss et al. prepared hb polystyrene via living anionic polymerization in the presence of vinylbenzyl chloride (VBC-a coupling

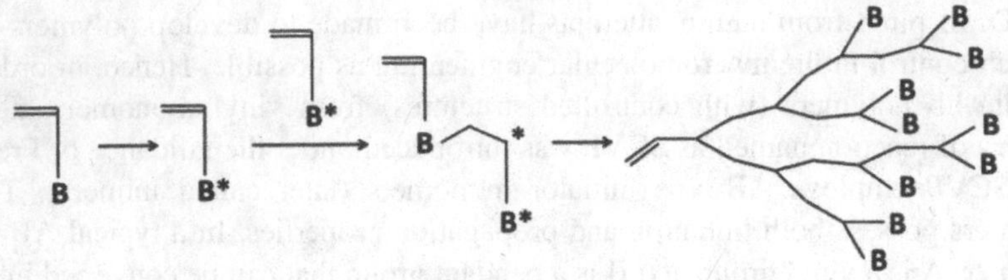

Scheme 3.9 Schematic representation of the synthesis of a hb polymer via SCVP of an AB-type monomer. Reprinted (adapted) with permission from Zhao [26]. Copyright (2015) Research Repository UCD (http://researchrepository.ucd.ie/handle/10197/6852)

agent) and nBuLi; Scheme 3.10 [30]. The synthesis procedure proceeded by slow addition of a chain-growth promoter (constituting of a polymerizable vinyl group and a functionality capable of quantitatively coupling with the living chain end). The addition of a chain-growth promoter to a living chain forms macromonomer which subsequently undergoes addition reactions to generate hb polymers. In another work, Baskaran reported synthesis of a hb polymer via living anionic SCVCP of a vinyl monomer, 1,3-diisopropenylbenzene (in the presence of nBuLi the vinyl monomer formed an AB-type monomer in situ) and a chain-growth promoter, styrene [31]. He observed that addition of styrene in the intermittent stages, during the polymerization reaction, increased M.W of the resultant hb polymer and decreased Mark–Houwink parameter, α significantly. In fact, low α values indicate densely packed three-dimensional structures. Recently, Young et al. developed vinyl functionalized hb polymers via living anionic SCVCP of allyl methacrylate and hydroxyethyl methacrylate. They showed that the presence of free vinyl pendant groups along the polymer backbone could be further functionalized according to needs (say for biomedical applications). Examples of living anionic SCVCP are rare due to instability of the anionic AB-type monomers. The method of slow addition of a stoichiometric quantity of a chain-growth promoter to the living polymer chain facilitates anionic SCVCP in few cases. Another disadvantage of anionic SCVCP is that the polymerization has to be carried out at sub-ambient temperature (below 0 °C). However, anionic SCVP/SCVCP is still considered as a challenging approach. In another variation, GTP technique as introduced by Dupont in the year 1983 is considered as one of the startling procedures to develop architectural polymers (block, graft, hb polymers etc.). GTP uses a trimethylsilyl ketene acetal initiator which is catalyzed by metal free nucleophilic anions. GTP is basically a "quasi-living" oxyanionic polymerization technique (generally follows Mukaiyama–Michael addition reactions), used for the controlled polymerization of α, ß-unsaturated carbonyl compounds. Mukaiyama–Michael addition reaction is a carbon–carbon coupling reaction which ensures high stereo selective control. GTP proceeds by repetitive transfer of a reactive silyl group from a propagating chain end to the monomer and thus produces a living end. Though, in later studies, it was proven that this transfer mechanism is inappropriate. According to new findings, silyl ketene acetals and a Lewis acid/base catalyst should coexist in the reaction

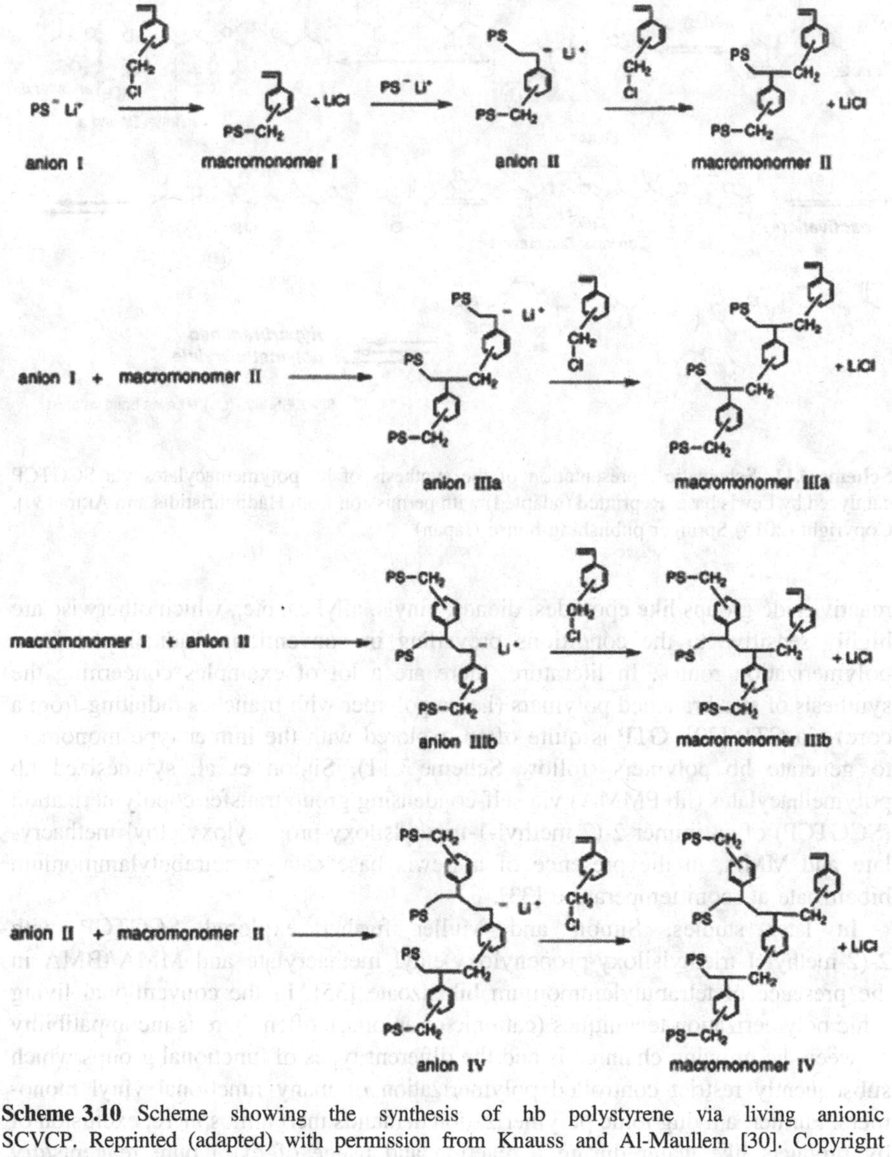

Scheme 3.10 Scheme showing the synthesis of hb polystyrene via living anionic SCVCP. Reprinted (adapted) with permission from Knauss and Al-Maullem [30]. Copyright (2000) Wiley online library

medium which actually features group transfer reactions. GTP produces living polymers with controlled topologies/narrow M.W.D/high DB, free of termination or chain transfer reactions just at room temperature and under mild reaction conditions. Recombination terminations are avoided in GTP as trialkylsilyl capped propagating chain ends are electrically neutral. Due to the suppression of undue side reactions, GTP is favored to polymerize functional methacrylates which bear

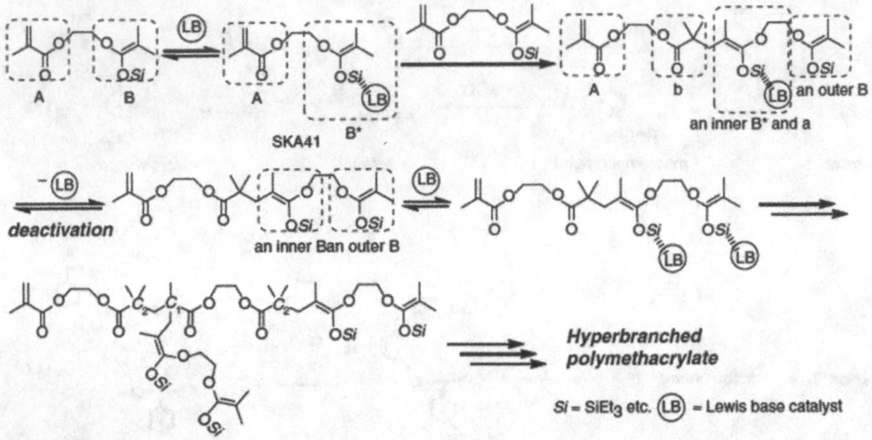

Scheme 3.11 Schematic representation of the synthesis of hb polymethacylates via SCGTCP catalyzed by Lewis base. Reprinted (adapted) with permission from Hadjichristidis and Akira [34]. Copyright (2015) Springer publishing house (Japan)

reactive side groups like epoxides, dienes, vinyls, allylics, etc., which otherwise are highly sensitive to the conditions prevailing in conventional radical and ionic polymerization routes. In literature, there are a lot of examples concerning the synthesis of star-branched polymers (i.e., a polymer with branches radiating from a core) via GTP [32]. GTP is quite often explored with the inimer type monomers to generate hb polymers (follow Scheme 3.11). Simon et al. synthesized hb polymethacylates (hb PMMA) via self-condensing group transfer copolymerization (SCGTCP) of an inimer 2-(2-methyl-1-triethylsiloxy-propenyloxy)ethyl methacrylate and MMA, in the presence of a Lewis base catalyst tetrabutylammonium bibenzoate at room temperature [33].

In later studies, Simon and Muller further explored SCGTCP with 2-(2-methyl-1-triethylsiloxy-propenyloxy)ethyl methacrylate and MMA/tBMA in the presence of tetrabutylammonium bibenzoate [35]. In the conventional living ionic polymerization techniques (cationic or anionic), often there is incompatibility between the growing chain ends and the different types of functional groups which subsequently restrict controlled polymerization of many functional vinyl monomers. Further, a living ionic polymerization demands inert atmosphere, exclusion of by products like water during a reaction and usage of extra pure reagents/dry solvents. These complications of living ionic polymerizations restrict their applications in many instances. Hence, it was necessary to develop polymerization techniques which combine the desirable features of both conventional FRP and ionic polymerization. Attempts to develop living radical polymerization systems in the earlier years included usage of iniferters (initiator-transfer agent-terminator) [36]. The basic concept of this approach was reversible termination of the growing polymer chains. However, polymerization carried out in the presence of iniferters yield polymers with broad M.W.D and poor properties (due to very low M.W). In the later years, while searching for various radical trappers (for studying the detailed chemistry of initiation in radical polymerization), nitroxides (like TEMPO and its derivatives) were found to combine with carbon centered radicals to form alkoxyamines and thus were considered as excellent initiator inhibitors. However, nitroxides either do not react or react reversibly with oxygen centers (tert-butoxy, benzoyloxy, etc.). Alkoxyamines (C–ON bonds) are thermally labile (typically at 80 °C or above) and undergo reversible dissociation. This concept was utilized to give birth to a new type of controlled radical polymerization technique based on alkoxyamine chemistry. In fact, with the further advancement in the research of living polymerization techniques, NMP first came into existence as a breakthrough.

The term NMP was introduced by an eminent scientist Ezio Rizzardo of Commonwealth Scientific and Industrial Research Organization, Australia. Moad from Rizzardo's group introduced the concept of reversible end capping of the propagating chain ends by TEMPO [38]. Typically, during NMP, at first C–ON bonds are formed at the end of the propagating polymer chains which break homolytically under the polymerization condition (in the following steps) to regenerate free nitroxide radicals and polymeric radicals. The polymer radicals extend the growth of the chains through normal propagation reactions and thus DP increases. Again, the polymer radicals (with higher DP) recombine with nitroxide radicals. This cycle of homolysis/monomer addition/recombinations repeats to maintain a living nature of the growing polymer chains. At any instance of NMP, there is a significant presence of inactive polymer chain ends, which substantially decrease the overall concentration of the reactive polymer chain ends and thus reduce the unwanted side reactions like bimolecular radical-radical termination, disproportionation, combination, or cyclic reactions. Moreover, the rate of cross reactions between nitroxide radicals and carbon centered radicals are much faster than the homo coupling reactions between the consecutive carbon-centered radicals [39]. These phenomena of "persistent radical effects" (PRE) enable the polymer chains to grow in a controlled manner. The equilibrium step in NMP is shown in Scheme 3.12. In the realm of hb polymers, NMP has been mostly explored through SCVP techniques. The AB-type monomer that is used in SCVP-based NMP constituted of a polymerizable double bond (A) and an initiating alkoxyamine group (B). In a novel work, Hawker et al. homopolymerized an AB-type styrenic monomer functionalized with an alkoxyamine initiating group, **9** at 130 °C in 72 h, to a hb polymer **10** without any gelation; Scheme 3.13 [40]. The resultant hb polymer exhibited a M.W around 6000 g mol^{-1} and a P.D.I of 1.4. Interestingly, they further utilized the hb polymer as a macroinitiator for a second step chain

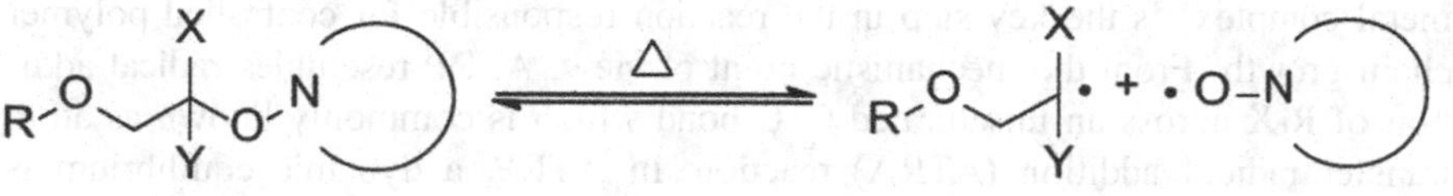

Scheme 3.12 Schematic representation of activation–deactivation equilibrium in NMP. Reprinted (adapted) with permission from Gigmes [37]. Copyright (2015) Royal society of chemistry

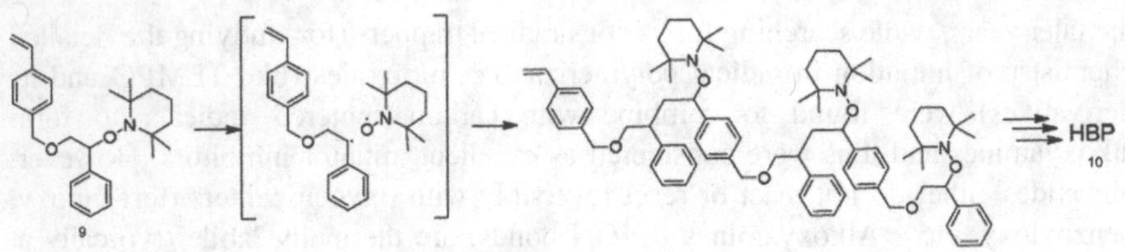

Scheme 3.13 Scheme showing homopolymerization of an alkoxyamine-based inimer via NMP-SCVP to generate a hb polymer. Reprinted (adapted) with permission from Wang and Gao [40, 41]. Copyright (2017) MDPI

extension to develop a hyperbranched-star (hyperstar) polymer with M.W around 3×10^6 g mol^{-1} and a P.D.I of 4.35.

Khan et al. discovered that the presence of pendant polymerizable groups along the polymer backbone may be useful in generating resist materials for UV-imprint lithography; UV-IL applications [42]. The group prepared different hb polymers from a variety of mono- and di-functional monomers using NMP. One of the hb polymer grades was prepared by copolymerization of styrene and a di-functional methacrylate (1,6-hexanediol dimethacrylate) mediated by TIPNO. Using this concept, NMP may be extended to generate hb polymers suitable for biomedical applications just by functionalization. Unfortunately, there is limited reports on the synthesis of hb polymers via NMP-SCVP due to slow polymerization kinetics that require high polymerization temperatures (often 120–145 °C)/lengthy polymerization duration (24–72 h), incompatibility of nitroxides with various functional groups of the vinyl monomers (say in case of MMA) and multistep synthesis of alkoxyamine-based inimers [39]. Additional disadvantages of NMP include homolysis of N–O bonds instead of C–ON bonds, chain transfer to solvent, loss of nitroxide radicals, oxidation of alkyne bearing monomers and high costs of the nitroxide initiators [39]. To overcome these shortcomings, researchers are altering the structures of nitroxides [43]. Unlike initiating radicals, nitroxide radicals are mediating radicals that are thoroughly involved in several activation/deactivation steps. Hence, the structure of the nitroxides definitely puts a significant effect on the polymerization. However, structurally altered nitroxides have still not been used to generate hb polymers and this leaves enough scope for exploration. Rather, other living polymerization techniques like ATRP and RAFT were developed to prepare hb polymers with controlled topologies and properties.

In recent times, ATRP is considered as one of the dominating polymerization techniques to prepare polymers with predictable structures and highly targeted M.W ranges. As the name suggests, atom transfer (from an organic halide to a transition metal complex) is the key step in the reaction responsible for controlled polymer chain growth. From the mechanistic point of view, ATRP resembles radical addition of R-X across an unsaturated C–C bond which is commonly known as atom transfer radical addition (ATRA) reaction. In ATRP, a dynamic equilibrium is maintained between the propagating radicals (R$^{\cdot}$ or P$_n$) and the dormant species (alkyl halides or pseudo halides; R-X or dormant propagating chain ends;

R-P_n-X) [45]. Typically, in ATRP, radicals are generated from a halide through a reversible redox reaction, catalyzed by a redox-active, low oxidation state transition metal complex (M_t^z-L_n where M_t^z denotes the metal atom or ion in oxidation state z and L_n denotes a ligand in the counter ion form). In this activation step, alkyl halogen bond (R-X) is reversibly and homolytically cleaved by M_t^z (activator, in the lower oxidation state) to generate an alkyl radical (R˙) and $M_t^{(z+1)}$ (deactivator, in the higher oxidation state) via an inner sphere electron transfer process. In the subsequent step, R˙ either attacks vinyl monomers to generate $P_n^˙$ or is reversibly deactivated by X-$M_t^{(z+1)}L_n$. X-$M_t^{(z+1)}L_n$ being a deactivator cleaves heterolytically, then rapidly transfers the X group back to the radicals and transform the radicals into dormant species (P_n-X). Actually, transition metal complexes are highly effective in transferring halogens than R-X. Activation and deactivation steps repeat throughout the polymerization reaction until all the vinyl monomers are consumed. Termination reactions occur by radical coupling and disproportionation. However, the termination step is suppressed to a minimum, but not eradicated, in a well-controlled ATRP. Due to PRE, when the polymerization progresses, the termination step is slowed down significantly as at any instance, the equilibrium is shifted towards the dormant species, i.e., $k_{act} \ll k_{deact}$ (where k_{act} and k_{deact} are the rate constants for the activation and deactivation steps, respectively). Mechanistically, X-$M_t^{(z+1)}L_n$ accumulates due to PRE and maintains a very low concentration of the propagating radicals. This phenomenon reduces the probability of bimolecular termination of the propagating radicals and prolongs the propagation step which eventually produces polymers with high M.W and uniform M.W.D. And also, topology of the polymer is well controlled in ATRP due to the absence of unwanted side reactions. The entire mechanism of ATRP is shown schematically in Scheme 3.14. So far, the most commonly used transition metal complex in protic medium ATRP is those of copper (due to two oxidation states Cu(I) and Cu(II)). The catalyst consists of Cu(I) halide accompanied by a nitrogen donor based complexing ligand. The ligands play a key role in solubilizing the catalysts and determining the redox potential of the catalysts which eventually guide the shifting of the equilibrium in ATRP. The synthesis of hb polymers using ATRP is often extended with SCVP [46]. For ATRP-SCVP, B group must carry a halogen atom that is capable of initiating through a reaction with the copper catalyst.

It was for the first time Gaynor et al. developed a hb polymer **12** from p-(chloromethyl) styrene (**11**, an AB-type monomer) via ATRP-SCVP; Scheme 3.15. In the following year, Matyjaszewski et al. synthesized hb polyacrylates via ATRP-SCVP of an AB-type monomer, 2-((2-bromopropionyl) oxy)ethyl acrylate (BPEA) [49]. Interestingly, Muthukrishnan et al. synthesized hb glycopolymers via ATRP-SCVCP of BPEA, **13** and a sugar carrying acrylate, 3-O-acryloyl-1,2,5,6-di-O-isopropylidene-α-D-glucofuranoside (AIG1c)- **14**, followed by deprotection of isopropylidene protecting groups in the final step; Scheme 3.16 [50]. These polymers have potential prospects in biological, pharmaceutical, and medical applications due to multiple binding sites.

In most cases, three types of AB-type monomer based on polymerizable vinyl groups are explored in ATRP-SCVP which includes acrylate inimers, methacrylate

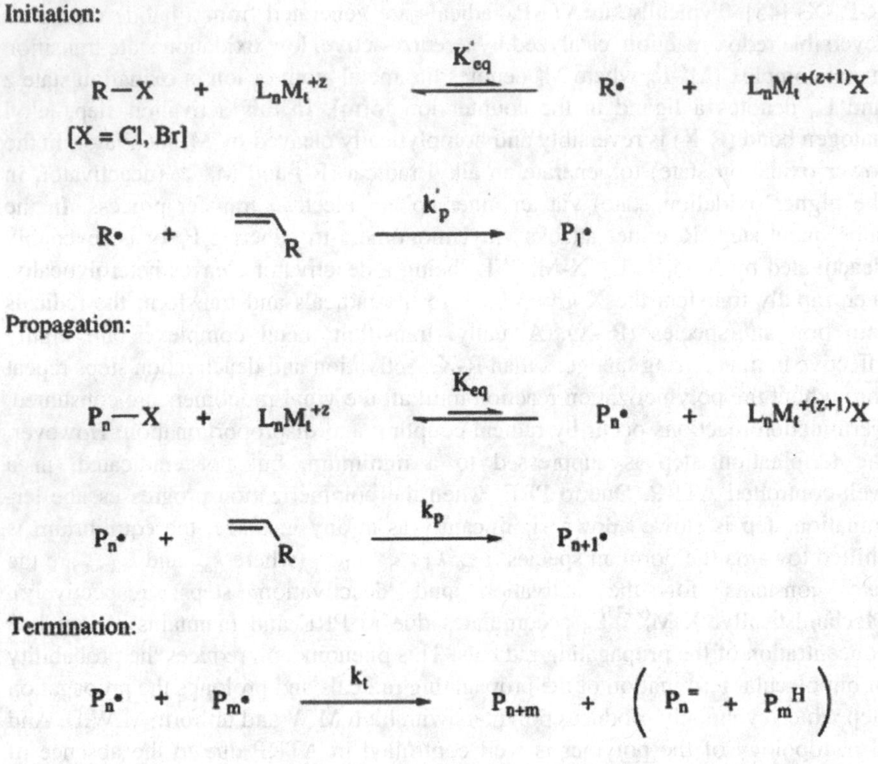

Initiation:

$$R-X \;+\; L_nM_t^{+z} \quad \underset{}{\overset{K_{eq}^{'}}{\rightleftharpoons}} \quad R\bullet \;+\; L_nM_t^{+(z+1)}X$$

$$[X = Cl, Br]$$

$$R\bullet \;+\; \underset{R}{\diagup\!\!\!=} \quad \overset{k_p^{'}}{\longrightarrow} \quad P_1\bullet$$

Propagation:

$$P_n-X \;+\; L_nM_t^{+z} \quad \underset{}{\overset{K_{eq}}{\rightleftharpoons}} \quad P_n\bullet \;+\; L_nM_t^{+(z+1)}X$$

$$P_n\bullet \;+\; \underset{R}{\diagup\!\!\!=} \quad \overset{k_p}{\longrightarrow} \quad P_{n+1}\bullet$$

Termination:

$$P_n\bullet \;+\; P_m\bullet \quad \overset{k_t}{\longrightarrow} \quad P_{n+m} \;+\; \left(P_n^{=} \;+\; P_m^{H} \right)$$

Scheme 3.14 Schematic representation of the mechanism of ATRP. Reprinted (adapted) with permission from Patten and Matyjaszewski [44]. Copyright (1998) Wiley online library

inimers, and styrenic inimers. Amin and Gaffar synthesized hb polyamides, **16** via ATRP-SCVP of an AB-type monomer, **15** designed from p-amino .phenol; Scheme 3.17 [51]. Tsarevsky et al. prepared degradable hb polymers with multiple alkyl halide chain ends via ATRP-SCVP of inimers containing esters (2-(2′-bromopropionyloxy)ethyl acrylate) or disulphides (2-(2′-bromoisobutyr-yloxy)ethyl 2″-methacryloyloxyethyl disulfide) groups [52]. It is worthy to mention that degradable polymers are highly demanded in biomedical applications.

The major disadvantage of ATRP in the design of hb polymers for successful biomedical applications is the usage of copper catalysts in high concentrations ($\sim$0.1–1 mol% with respect to the monomers). One cannot reduce the amount of Cu species in the initial recipe because the bimolecular termination reactions consume activators and the polymerization does not proceed to completion. Removal of copper contaminants from the system is difficult and expensive (say by extraction, precipitation, immobilization, or by using biphasic systems) [53]. However, a lot of efforts have been put to reduce copper contamination in order to prevent ATRP from dying. Usage of iron catalysts instead of copper catalysts is often favored in biomedical applications due to low toxicity and biocompatibilty.

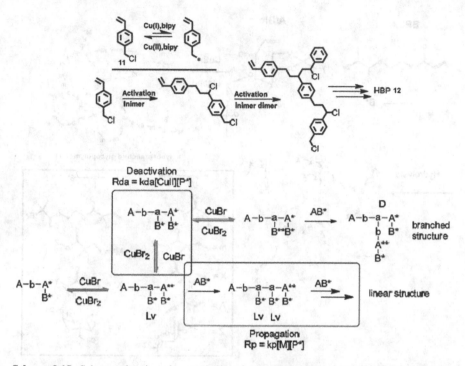

Scheme 3.15 Scheme showing the synthesis of hb polystyrene via ATRP-SCVP from an AB-type monomer. A* and B* represent two types of initiating groups whereas A** and B** represent two types of propagating radicals. Reprinted (adapted) with permission from Wang and Gao [41, 47, 48]. Copyright (2017) MDPI

However, iron catalysts are not so efficient like copper catalysts and are quite expensive. An effort was made to diminish the concentration of copper catalysts by the employment of activators regenerated by electron transfer (ARGET) process [53]. Typically in ARGET process, a reducing agent like Sn(II) 2-ethylhexanoate; Sn(EH)$_2$, ascorbic acid, or sugars like glucose (which are generally FDA approved) reduce the accumulating Cu(II) species by transforming them to Cu(I) species. This continuously regenerates activators and thus reduces the catalyst concentration in the initial recipe. In fact, ARGET process can reduce the catalyst concentration in an ATRP recipe to 50 ppm [54]. Interestingly, these redox processes do not generate initiating radicals or initiating species and thus pure copolymers can be synthesized. Another effort is the employment of initiators for continuous activator regeneration (ICAR) process in ATRP-SCVP. Typically, in an ICAR process, conventional initiator radicals (say AIBN) reduce the accumulating Cu(II) species to Cu(I) species. This process definitely diminishes copper loading and avoids the addition of external reducing agents. However, care should be taken that the decomposition of the selected free radical initiator must be sufficiently low in order to discourage bimolecular termination. Both ARGET and ICAR processes leave enough scope for exploration to produce biologically friendly hb polymers by

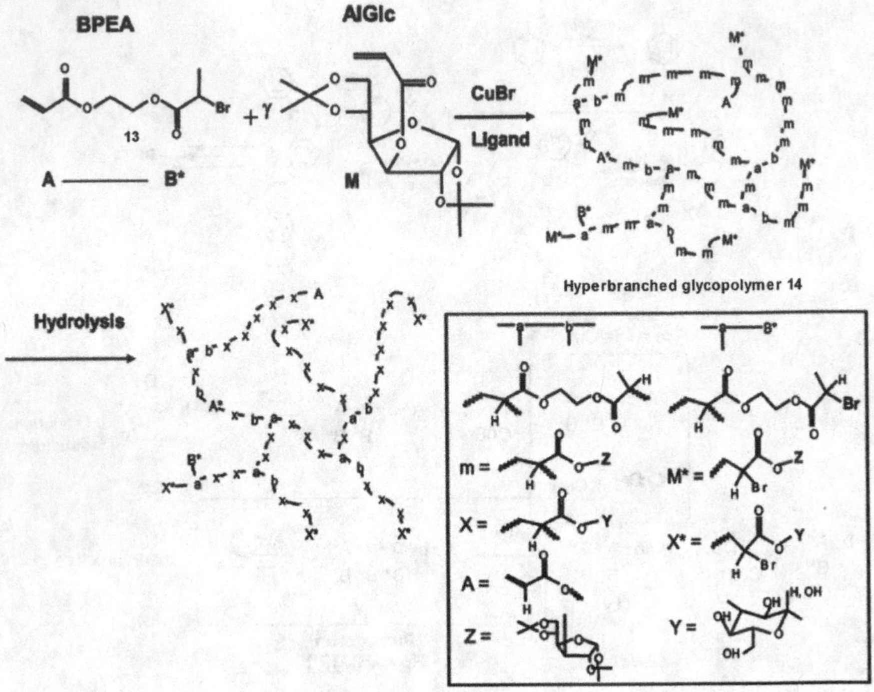

Scheme 3.16 Scheme showing the synthesis of hb glycopolymers via ATRP-SCVCP. Reprinted (adapted) with permission from Muthukrishnan et al. [50]. Copyright (2005) American chemical society

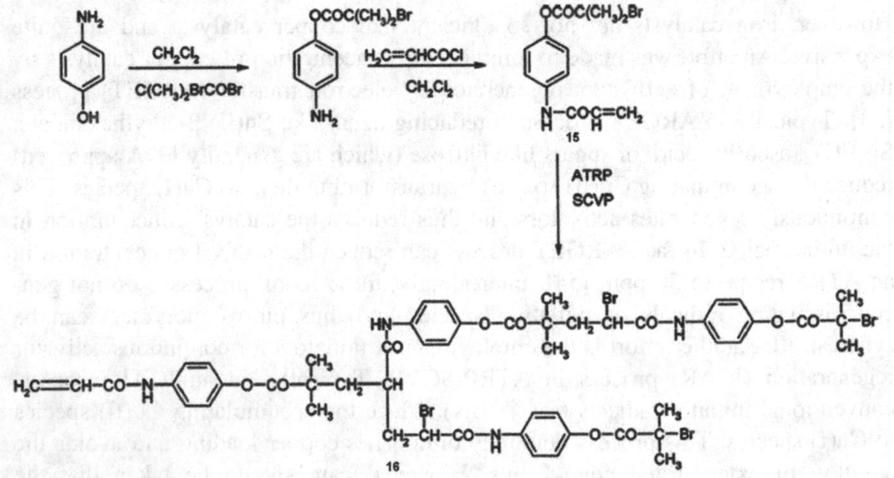

Scheme 3.17 Scheme showing the synthesis of hb polyamide via ATRP-SCVP. Reprinted (adapted) with permission from Amin and Gaffar [51]. Copyright (2008) Taylor and Francis online

ATRP for commercialization [55]. With the advancement in ATRP, in the most updated cases, in the synthesis of hb polymers, Cu(0) mediated ATRP has also gained impetus. For the first time, Matyjaszewski et al. synthesized hb polyacrylates via Cu(0) mediated ATRP-SCVP [56]. Actually, in SCVP, due to high initiator concentration, often a shift in the equilibrium towards the active radicals is favored. This increases the concentration of Cu(II) deactivator species in the system which eventually prevents polymerization. Cu(0) facilitates reduction of Cu(II) to Cu(I). Cu(I) species generated by Cu(0) are highly active and thus reacts with ATRP initiators so rapidly that the disproportionation to regenerate Cu(0) and Cu(II) species gets suppressed significantly. This approach is indirectly used to activate dormant halide (pseudo halide) species via an inner sphere electron transfer process and is thus named as supplementary activators and reducing agents ATRP (SARA-ATRP); Scheme 3.18. SARA-ATRP also diminishes amounts of the copper catalysts in the initial recipe and improves oxygen tolerance of the catalysts which is often a critical issue in the commercialization of ATRP. To address the issues of toxicity of copper catalysts, Ag(0) and Fe(0) have also been implemented in SARA-ATRP [55]. The mechanism as explained in SARA-ATRP is often a controversial issue. There is a complementary mechanism in Cu(0) mediated ATRP named as single electron transfer living radical polymerization (SET-LRP); Scheme 3.18. In SET-LRP, Cu(0) instead of Cu(I) single-handedly activates dormant halide (pseudo halide) species via an outer sphere electron transfer process.

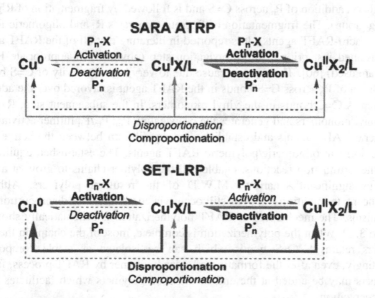

Scheme 3.18 Schematic representation of SARA-ATRP and SET-LRP. *Bold arrows* indicate major reactions, *solid arrows* indicate supplementary reactions, and *dashed arrows* indicate minor reactions. Reprinted (adapted) with permission from Konkolewicz et al. [57]. Copyright (2014) Royal society of chemistry

At any instance during a polymerization, Cu(I) spontaneously disproportionates into Cu(0) and Cu(II) as the reaction has a very low activation energy.

Xue et al. synthesized hb poly (methyl acrylate)-block-poly (acrylic acid)s; HBPMA-b-PAAs via SET-LRP technique [58]. HBPMA-b-PAAs spontaneously formed unimolecular micelles, constituting of hydrophobic cores (PMA) and hydrophilic shells (PAA), in an aqueous environment above pH 3. However, HBPMA-b-PAAs formed regular quadrangular prisms in an aqueous medium having a pH less than 2. Undoubtedly, such pH responsive HBPMA-b-PAAs may prove to be potential candidates for the delivery devices of biomacromolecules.

Finally, another widely exploiting living polymerization technique is RAFT, which is again highly favored in the design of hb polymers with controlled architectures. RAFT is easy to explore due to high tolerance of the reactants to various functional monomers (vinyl acetate, N-vinyl pyrrolidone etc.) and mild reaction conditions (ambient temperature and tolerance to oxygen). Typically, RAFT polymerization proceeds via reversible, degenerative (addition fragmentation) chain transfer processes, unlike in conventional chain transfer processes in FRP where chain transfer occurs only once. CTAs employed in RAFT polymerization are commonly named as RAFT agents like thiocarbonates, thiocarbamates, or dithioesters, all of which have a structure $Z(C=S)SR$. The RAFT agents maintain equilibrium between the active and the dormant species. In the first step (i.e., initiation step), free radicals are generated which attacks vinyl monomers and generate primary polymer radicals ($P_n^{\cdot}$). Just at the start of propagation, $P_n^{\cdot}$ react with the RAFT agents through chain transfer reactions. The chain transfer reaction proceeds via addition of $P_n^{\cdot}$ across $C=S$ and is followed by fragmentation of R group (leaving group). The fragmentation of R group generates $R^{\cdot}$ and oligomeric RAFT agents (macro-RAFT agents). It is reported in literature that all of the RAFT agents are consumed just prior to the propagation step (obviously if appropriate RAFT agents are used) [60]. This is so because due to very high reactivity of $C=S$ bonds, the addition of $P_n^{\cdot}$ across $C=S$ bonds in the RAFT agent is favored over the addition of $P_n^{\cdot}$ across $C=C$ bonds in the vinyl monomers. In the subsequent step, $R^{\cdot}$ reinitiates vinyl monomers and generates the propagating $P_m^{\cdot}$. $P_m^{\cdot}/P_n^{\cdot}$ further activates the oligomeric RAFT agents and establishes an equilibrium between the active $P_n^{\cdot}/P_m^{\cdot}$ and the dormant oligomeric/polymeric RAFT agents. The established equilibrium limits the termination reactions, enables all the polymer chains to grow at a time, and thus significantly narrows M.W.D of the resultant polymers. Although insignificant, termination reactions still occur via combination or disproportionation mechanisms. The mechanism of RAFT polymerization is schematically shown in Scheme 3.19. When the polymerization is complete, most of the chains in the dead polymers retains S–C=S bonds which can be isolated as stable compounds. Interestingly, even after the formation of a dead polymer by RAFT process, further monomers may be added at the end of the dead polymers which facilitates block copolymerization.

For the design of hb polymers, RAFT-SCVP is highly popular. In RAFT-SCVP, the AB-type monomer is basically a RAFT agent containing a polymerizable vinyl group and is named as transmer [41, 61]. Unlike the inimers used in NMP-SCVP or

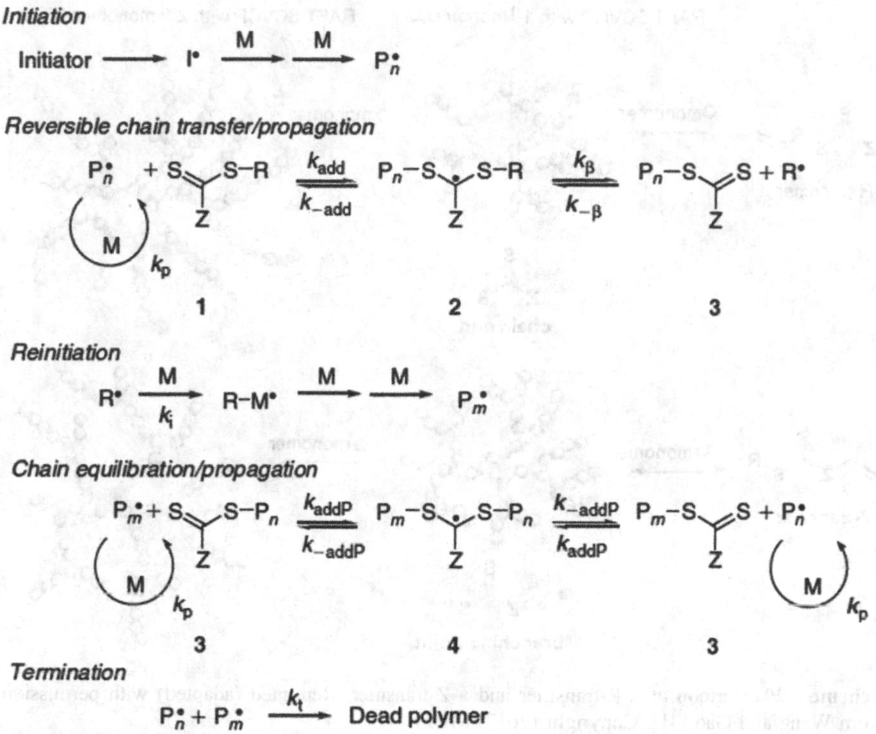

Initiation

$$\text{Initiator} \longrightarrow \text{I}^\bullet \xrightarrow{\text{M}} \xrightarrow{\text{M}} P_n^\bullet$$

Reversible chain transfer/propagation

      1   **2**   **3**

Reinitiation

$$R^\bullet \xrightarrow[k_i]{\text{M}} R\text{-}M^\bullet \xrightarrow{\text{M}} \xrightarrow{\text{M}} P_m^\bullet$$

Chain equilibration/propagation

      3   **4**   **3**

Termination

$$P_n^\bullet + P_m^\bullet \xrightarrow{k_t} \text{Dead polymer}$$

Scheme 3.19 Schematic representation of the mechanism of RAFT polymerization. Reprinted (adapted) with permission from Moad et al. [59]. Copyright (2006) CSIRO (Australia)

ATRP-SCVP, B groups in the inimers for RAFT-SCVP have to be initiated by R˙ which in turn is cleaved by an external radical initiator. A transmer may be designed in two forms; the polymerizable vinyl group may be present at the R group (R approach) or at the Z group (Z approach); Scheme 3.20. The R approach is more popular than the Z approach because in the latter approach, steric hindrance restricts access to CTA functionalities and the generated branch points are very weak. Actually in the Z approach, the initiating thiocarbonylthio groups are located at the branch points of the hb polymers which increases steric hindrance and eventually results in hydrolytically unstable groups at every branch point. On the contrary, in the R approach, the thiocarbonylthio groups are present at the polymer chain ends and thus results in highly peripheral functional polymers. The first work on RAFT-SCVP was carried out by Yang et al. where they synthesized a hb polymer from an AB-type monomer designed by introducing dithioester groups into the styrene monomers [62]. Unfortunately, this approach being a Z approach, the hb polymers exhibited weak branching points. Interestingly, in a recent work, Kalourkoti et al. synthesized segmented amphiphilic hb block copolymers of styrene and 2-vinylpyridine or 4-vinylpyridine via RAFT-SCVP [63]. In this work, they at first synthesized a hb polymer utilizing one of the monomer. The hb

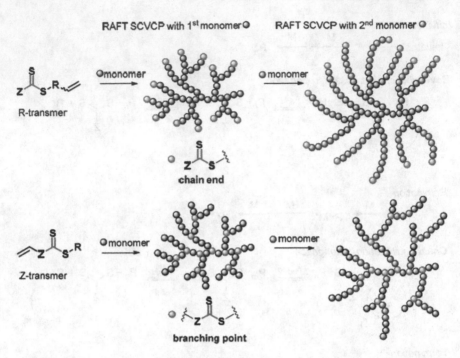

Scheme 3.20 Cartoon of a R-transmer and a Z-transmer. Reprinted (adapted) with permission from Wang and Gao [41]. Copyright (2017) MDPI

homopolymer contained thiocarbonylthio groups at the branch points which facilitated further polymerization with the second monomer. As the second monomer was inserted at the branch points, it resulted in hb block copolymers. The final amphiphilic hb block copolymer self-assembled into micelles with sizes greater than 10 nm. Heidenreich and Puskas synthesized a hb copolymer via RAFT-SCVP (R approach) of styrene and 4-vinylbenzene dithiobenzoate (an AB-type monomer) in bulk conditions (at 110 °C) [64]. The resultant hb copolymer exhibited a high M.W around 3.7×10^5 g mol^{-1} and a P.D.I of 2.65. In an interesting work, Carter from Rimmer's group prepared hb poly (*N*-isopropylacrylamide); hb PNIPAM (**18A/B/C**) via RAFT-SCVP like technique from NIPAM—**17A**, in the presence of a branching monomer that contained imidazole groups—**17B** which could be effectively transferred to the polymer chain ends through chain transfer reactions; Scheme 3.21 [65]. They found that the hb polymer was temperature responsive and was highly effective in purifying His-tagged BRCA-1 protein fragments by precipitation.

In an attempt to develop hb polymers with suitable properties for biological and biomedical applications, Ghosh Roy and De synthesized amino acid containing hb polymers, **20** via RAFT-SCVP approach from tert-butyl carbamate (Boc)-L-valine acryloyloxyethyl ester (Boc-Val-HEA), **19A** and S-(4-vinyl)benzyl

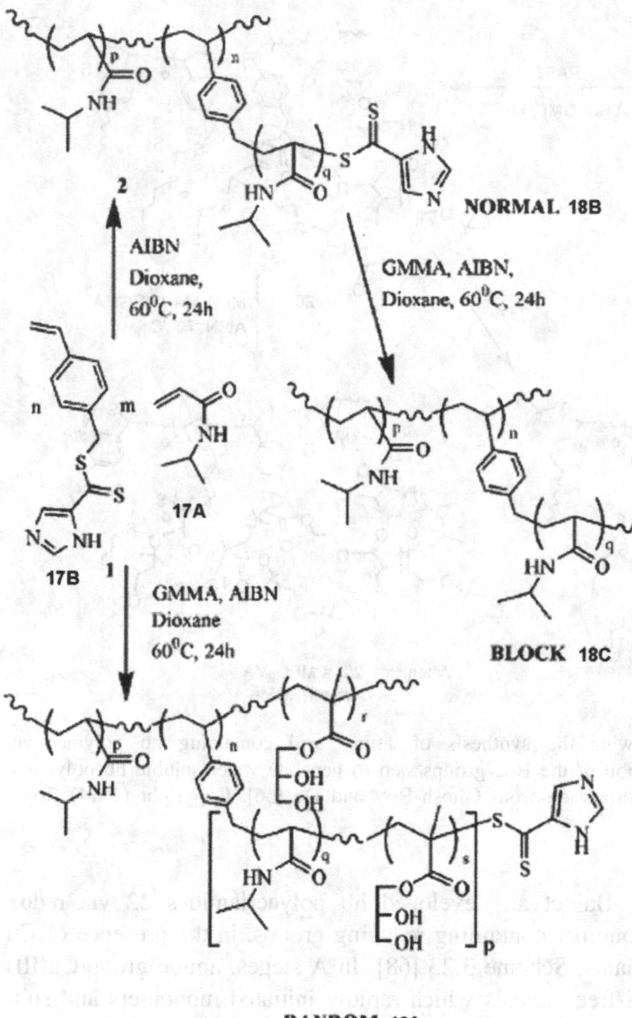

Scheme 3.21 Scheme showing the synthesis of hb PNIPAM rich in imidazole groups via RAFT-SCVP type approach. Reprinted (adapted) with permission from Carter et al. [65]. Copyright (2005) Wiley online library

S′-butyltrithiocarbonate (VBBT), **19B**; Scheme 3.22 [66]. When Boc was removed, water soluble, pH responsive, biocompatible, cationic hb polymers were obtained.

Han et al. prepared a series of chain segmented hb poly (tertiary amino methacrylate)s; HPTAMs with hydrophilic cores and hydrophobic shells, via RAFT-SCVP of 2-(dimethylamino)ethyl methacrylate (DMAEMA) and 2-((2-(((dodecylthiocarbonothioyl)thio)-2-methylpropanoyl)oxy)-ethyl acrylate (ACDT, an inimer) [67]. HBTAMs possessed regular linear chains between every two adjacent branching points for which they resembled HyperMacs in structures.

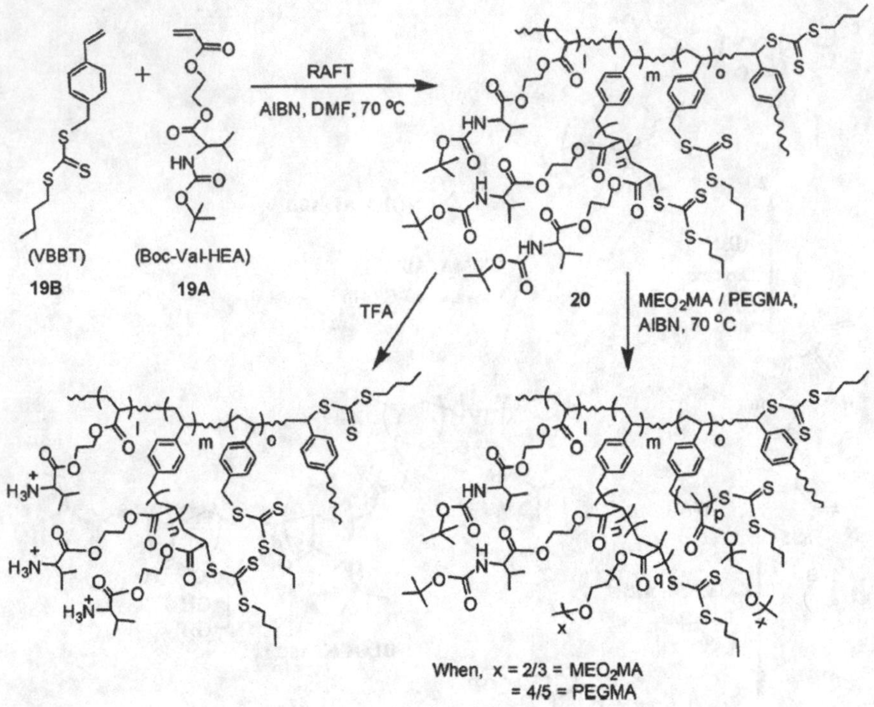

Scheme 3.22 Scheme showing the synthesis of amino acid containing hb polymer via RAFT-SCVP and deprotection of the Boc groups step to generate water soluble hb polymers. Reprinted (adapted) with permission from Ghosh Roy and De [66]. Copyright (2014) Royal society of chemistry

In an interesting work, Bai et al. developed hb polyacrlamides **22** via redox RAFT-SCVP from a monomer containing reducing groups, in the presence of Cu (III) and Ce(IV) as oxidants; Scheme 3.23 [68]. In A stages, amide group/Cu(III) redox process generated free radicals which rapidly initiated monomers and subsequently continued propagation. In B stages, in the excess of oxidizing agents, the linear polymers got further oxidized to form hb polymers.

Recently, the combination of RAFT and macromolecular architecture design via interchange of xanthates (MADIX) polymerization technique has gained much attraction to produce polymers with a variety of architectures and controlled properties. RAFT and MADIX follow the same mechanism but differ only by the mediator used. Delduc introduced the concept of degenerative chain transfer of radical species to xanthates; Scheme 3.24 [69]. Zhou et al. synthesized hb poly (vinyl acetate); Hb PVAc, **24** via RAFT/MADIX-SCVP from vinyl acetate, in the presence of vinyl 2-(ethoxycarbonothioylthio) acetate (ECTVA; a xanthate-based RAFT agent), **23**; Scheme 3.25 [70]. Hb PVAc **24** on hydrolysis produced hb PVA which is a potential pharmaceutical material.

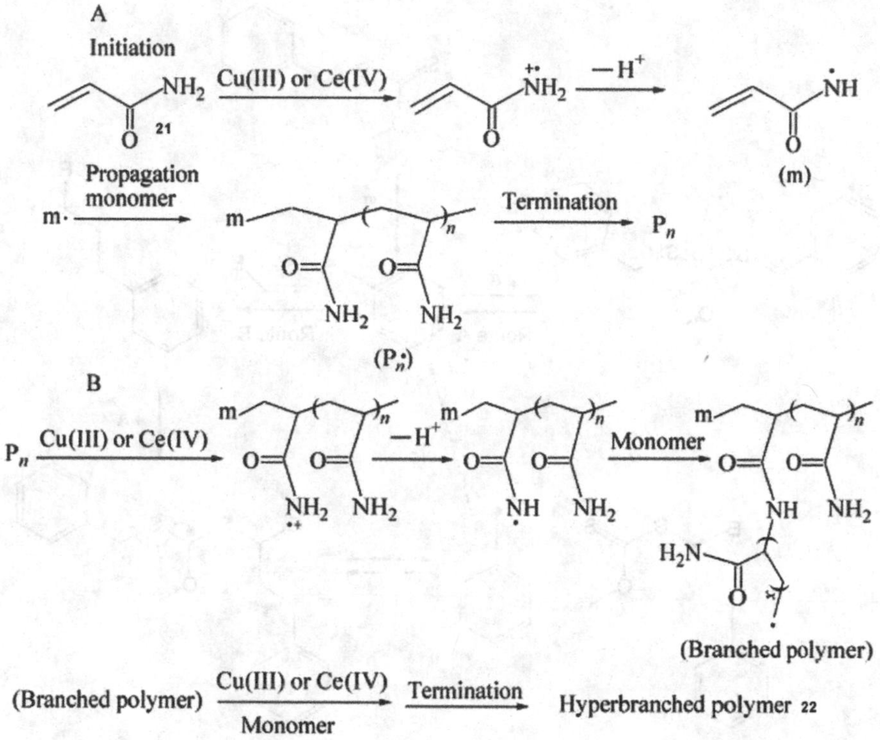

Scheme 3.23 Scheme showing the synthesis of a hb polymer via redox RAFT-SCVP. Reprinted (adapted) with permission from Bai et al. [68]. Copyright (2014) Springer

One of the disadvantages of RAFT polymerization is that the synthesized polymers are colored (either pink or yellow) due to the presence of S–C=S end groups. Hence, RAFT is often discouraged in commercial processes. There are certain techniques to remove the colorant groups from the polymers like aminolysis to produce thiol terminated polymers and transforming S–C=S groups to terminal hydrogen groups through a reaction with tri-n-butylstannane [72, 73]. Another shortcoming of RAFT-SCVP is that the synthesized hb copolymers exhibited higher M.W than the hb homopolymers. The preparation of hb polymers from cyclic AB-type monomers (cyclic inimers) relies on a new strategy named as ring opening multi-branching polymerization (ROMBP) or self-condensing ring opening polymerization (SCROP). In this approach, a strained, cyclic group generates a branching point only upon ring opening. ROMBP is considered as one of the important synthesis process to develop hb polymers for biomedical applications like hb polyamines, hb polyethers, hb polyesters, and hb poly siloxanes [74]. ROMBP may be further categorized as cationic ROMBP, anionic ROMBP, and catalytic ROMBP [75]. Hauser developed hb poly(ethyleneimine)s via cationic ROMBP of aziridines (three-membered alkylene imines) in the presence of cationic initiators [76]. Herein, intermolecular nucleophilic attack of the secondary amine

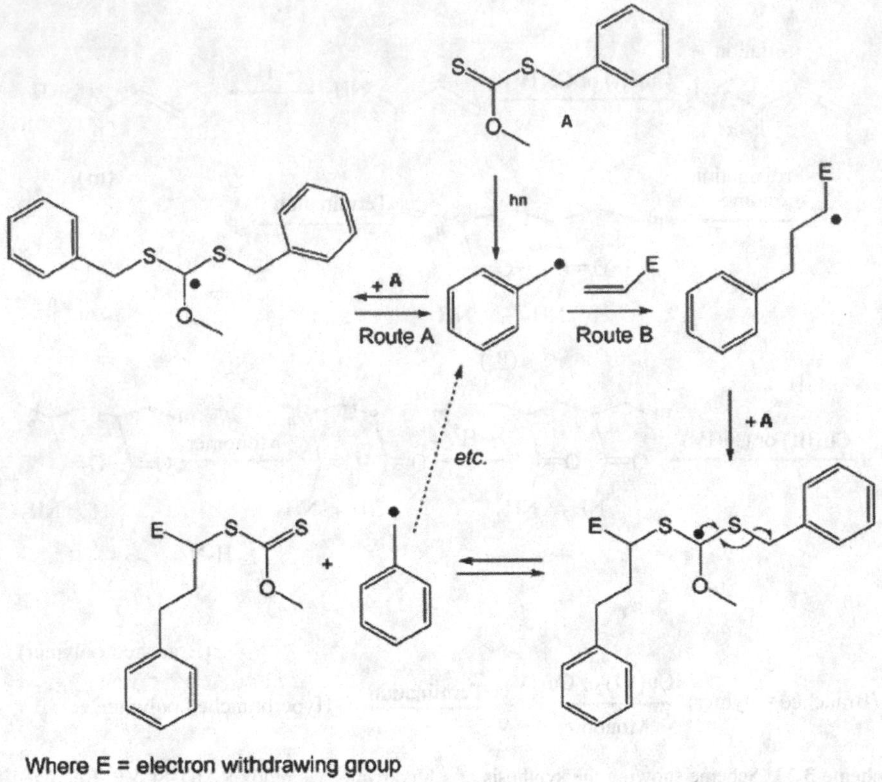

Where E = electron withdrawing group

Scheme 3.24 Schematic representation of degenerative transfer mechanism in the presence of xanthates. Reprinted (adapted) with permission from Perrier and Takolpuckdee [69, 71]. Copyright (2005) Wiley online library

nitrogens (present along the polymer backbone) on the propagating iminium centers produces tertiary amine groups which eventually cause branching; Scheme 3.26. Hb poly (ethyleneimine)s synthesized from aziridines is a well-known commercial product, sold under the trade name Lupasol®. Other cyclic monomers based on oxiranes and oxetanes have also been polymerized via cationic ROMBP.

Generally, cyclic amides and esters (lactams and lactones), cyclic ethers or Leuchs' anhydrides, and other vinyl monomers with electron withdrawing groups (like acrylonitrile, methyl vinyl ketone, etc.) are polymerized via anionic ROMBP. A typical anionic ROMBP mechanistic pathway is shown in Scheme 3.27. Among them, synthesis of hb polyglycerols (hbPGs) is considered as one of the important anionic ROMBP approaches. Vandenberg reported anionic ROMBP of glycidol using KOH as an initiator, for the first time [77]. Unfortunately, he obtained branched PG oligomers only. This happened because alkoxide anions tend to form aggregated species (through both inter- and intramolecular associations) in polar solvent which eventually caused a slow

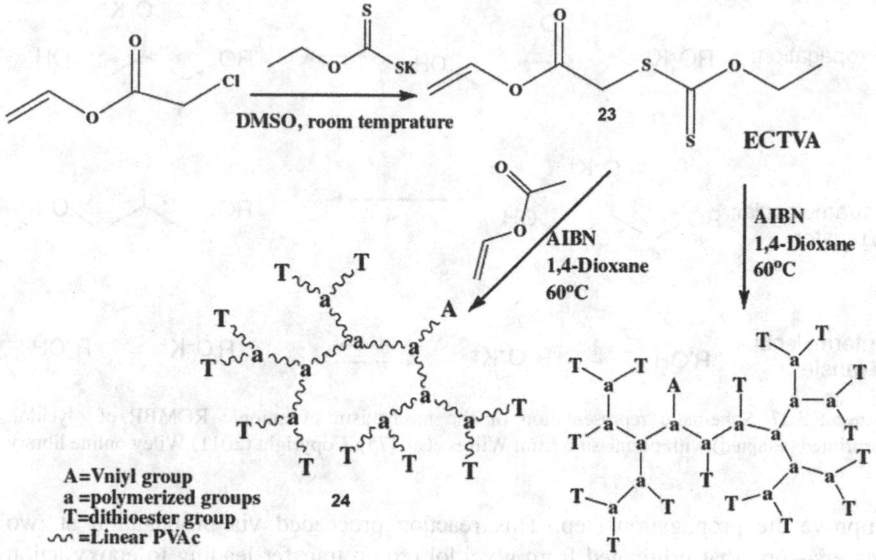

Scheme 3.25 Scheme showing the synthesis of hb PVAc via RAFT/MADIX-SCVP. Reprinted (adapted) with permission from Zhou et al. [70]. Copyright (2011) Elsevier

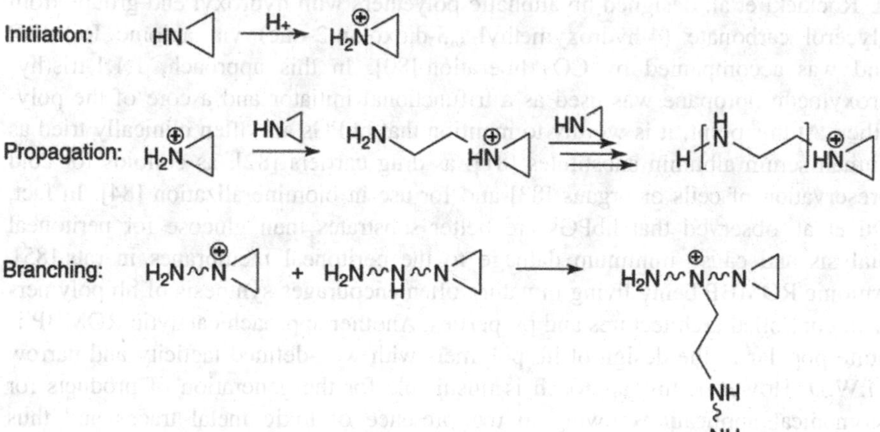

Scheme 3.26 Scheme for cationic ROMBP of ethylene imine. Reprinted (adapted) with permission from Wilms et al. [75]. Copyright (2011) Wiley online library

propagation. Sunder et al. improved the propagation in ROMBP of glycidol by slow monomer addition technique and obtained hbPGs with M.W around 6000 g mol^{-1} and P.D.I < 1.3 [78]. Goodwin and Baskaran prepared hbPGs via epoxy inimer mediated ROMBP where they used glycol as an initiator in the presence of potassium counter ion [79]. They employed batch monomer addition in order to

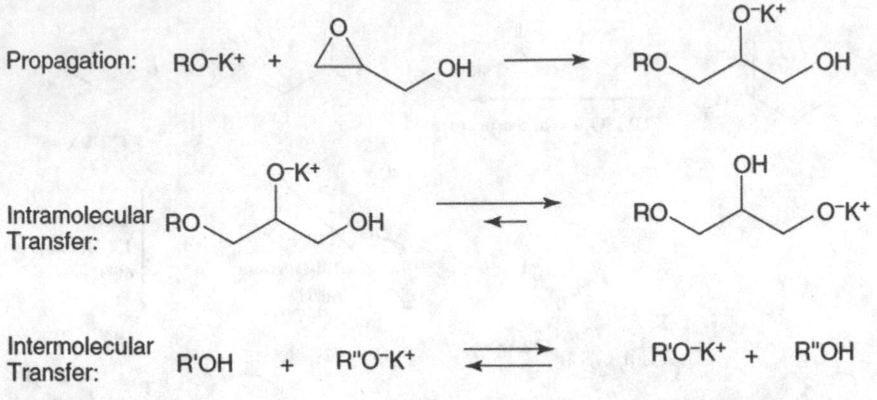

Scheme 3.27 Schematic representation of the mechanism of anionic ROMBP of glycidol. Reprinted (adapted) with permission from Wilms et al. [75]. Copyright (2011) Wiley online library

improve the propagation step. This reaction proceeded via propagation at two centers—one that originated from glycidol proton transfer leading to epoxy anion inimer and the other that undergoes hyperbranching without any transfer to monomers. In this approach, equilibrium was maintained between oligomers and high M.W hbPGs.

Rockicki et al. designed hb aliphatic polyethers with hydroxyl end groups from glycerol carbonate (4-hydroxymethyl-1,3-dioxolan-2-one) via anionic ROMBP and was accompanied by CO_2 liberation [80]. In this approach, 1,1,1-tris(hydroxymethyl)propane was used as a trifunctional initiator and a core of the polyether. At this point, it is worthy to mention that hbPGs are often clinically tried as human serum albumin substitutes [81], as drug carriers [82], as colloids for cold preservation of cells or organs [83] and for use in biomineralization [84]. In fact, Du et al. observed that hbPGs are better substrates than glucose for peritoneal dialysis and cause minimum damage to the peritoneal membranes in rats [85]. Anionic ROMBP being living in nature often encourages synthesis of hb polymers with controlled architectures and properties. Another approach catalytic ROMBP is quite popular in the design of hb polymers with well-defined tacticity and narrow M.W.D. However, this approach is unsuitable for the generation of products for biomedical applications owing to the presence of toxic metal traces and thus discouraged.

Both SCVP (normal or controlled) and ROMBP are highly acknowledged in the design of hb polymers but the availability of any type of inimer or their stringent preparation processes restricts industrialization. Often the slow monomer addition techniques are followed in SCVP/ROMBP to obtain hb polymers with higher DB. In future, further research has to be done in developing easier methods to synthesize inimers from the conventional vinyl monomers so that commercialization of SCVP and ROMBP approaches are possible.

3.5 Hypergrafting

Hypergrafting, i.e., covalent attachment of hb polymers to different substrates generates hybrid structures with complex architectures which is often favored in many biomedical applications [87]. There are two categories of hypergrafting—homogeneous hypergrafting (if the substrates are soluble like hydrophilic polymers) and heterogeneous hypergrafting (if the substrates are insoluble like metal nanoparticles or silicon wafers); Scheme 3.28.

3.5.1 Homogeneous Grafting—Hyperbranched-Graft-Hyperbranched Copolymers

When hb polymers are grafted from the surface of hb macroinitiators (like low M.W hbPGs) then hyperbranched-graft-hyperbranched copolymers (hb-g-hb) are generated. Hb-g-hb copolymers have core–shell structures which are highly suitable for

(a) Complex Polymer Architectures

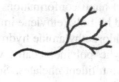

| Hyperbranched-Hyperbranched Graft Copolymers (HHGCs) | Linear-Hyperbranched Graft Copolymers (LHGCs) | Linear-Hyperbranched Block Copolymers (LHBCs) |

(b) Surface Functionalization

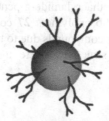

Spherical Particles   Planar Surfaces

Scheme 3.28 Cartoon showing various substrates available for hypergrafting; **a** homogeneous hypergrafting and **b** heterogeneous hypergrafting. Reprinted (adapted) with permission from Schull and Frey [86]. Copyright (2013) Elsevier biomedical transport applications [67, 88]. Xu et al. synthesized amphiphilic hb-g-hb copolymers via cationic ROMBP of 3-ethyl-3-hydromethyl oxetane (EHMO) and glycidol [89]. The resultant hb-g-hb copolymers were composed of hydrophobic Hb-PEHMO cores and hydrophilic HbPG shells. Popeney et al. developed a dendritic core–shell nanostructures via a two-step process—generation of nonpolar dendritic PE cores by late transition metal catalyzed chain walking polymerization which was followed by anionic ROMBP of glycidol to graft hbPG shells [88]. The core–shell nanoparticles exhibited high internal polarity gradient which facilitated the transport of poorly water soluble molecules. Unfortunately, there is very limited works on hb-g-hb copolymers and therefore it leaves enough scope for research.

3.5.2   Homogeneous Grafting— Linear-Graft-Hyperbranched Copolymers

When hb polymers are grafted from the surface of linear macroinitiators, then linear-graft-hyperbranched copolymers (lin-g-hb) are generated. Lin-g-hb copolymers have attracted much attention in the realm of macromolecules design because they have a high number of functional groups at the side chains and also exhibit cylindrical nano conformations in bulk or solution (just like DenPols). Kuo et al. developed hb poly (ethylene imine); hb PEI from poly (allyl amine) macroinitiators using 2-chloroethylamine hydrochloride [90]. The resultant polymer constituted of linear allylic polymer chains with pendant hb PEI and thus was considered as excellent multident chelates. Schull and Frey synthesized lin-g-hb copolymers of linear (4-hydroxy styrene)-graft-hyperbranched PGs (PHOS-g-hbPG), via three steps anionic and oxy-anionic polymerizations [91].

In an effort to increase M.Ws of the lin-g-hb copolymers, a grafting to strategy was developed; Scheme 3.29. Schull et al. prepared hbPGs dendron analogues, 25 containing exactly one focal amino group by ROMBP of glycidol in the presence of a trifunctional initiator (N,N-dibenzyl tris (hydroxylmethyl) amino methane) [92]. Then they attached hbPGs dendron analogues to poly (pentafluorophenol methacrylate), 26 to develop poly (hbPGs methacrylamide-g-pentafluorophenol methacrylate); hbPGMA-g-PFPMA, 27. Lin-g-hb cpolymers 27 could be further functionalized with drugs or biomedical imaging compounds due to the presence of multifunctionality of hbPG dendrons in the structures. Here again, we have obtained only few works on lin-g-hb polymers.

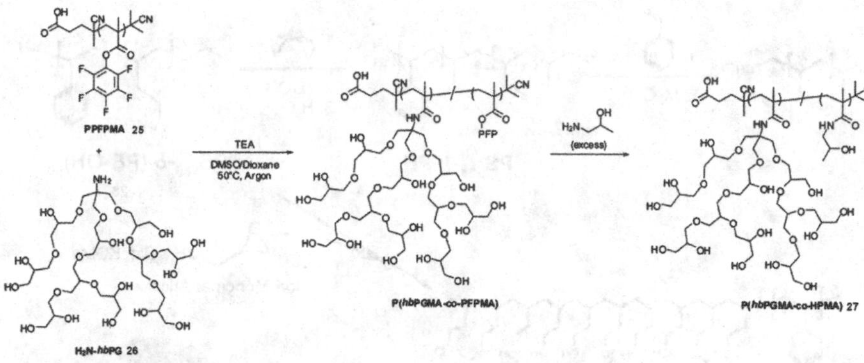

Scheme 3.29 Scheme showing the synthesis of lin-g-hb copolymer via grafting to strategy. Reprinted (adapted) with permission from Schull et al. [92]. Copyright (2012) American chemical society

3.5.3 Homogeneous Grafting—Linear-Block-Hyperbranched Copolymers

Linear-block-hyperbranched copolymer (lin-b-hb) is a block copolymer which consists of a hyperbranched/dendrimer block linked to the chain ends of a linear block. The hb block adds many functional groups to the linear blocks and thus develops polymers with many new interesting properties. There are three ways in which lin-b-hb copolymers may be synthesized: (1) "chain first"—hypergrafting from the linear segments which bear initiating groups for the generation of the hb blocks through polymerization, (2) "coupling strategy"—hypergrafting through the reactions between monofunctional hb blocks and end functional linear blocks and (3) "core first"—hypergrafting from the hb blocks (cores) which bears a single initiating group for divergent polymerization [93]. There are numerous examples of lin-b-hb copolymers due to the ease of synthesis.

Barriau et al. synthesized an amphiphilic block copolymer constituting of a linear apolar block and a hydrophilic hb block; Scheme 3.30 [94]. Initially, they prepared a diblock copolymer of polystyrene and linear hydroxylated polybutadiene (PS-b-PBOH), **28** via anionic polymerization and subsequent hydroboration reaction. PS-b-PBOH acted as a macroinitiator for glycidol (an AB_2-type monomer), in the post-polymerization grafting step which generated PS-b-PB-b-hbPG, **29**. Lin-b-hb copolymer **29** formed self-assembled micelles in various apolar solvents. These micelles appeared to be suitable for the generation of nano-reactors on the surfaces which may facilitate biomineralization and other biomedical applications. Marcos et al. used PS-b-PBOH macroinitiator for the grafting of hb poly (carbosilane); hb PCS blocks from methyldiundec-10-enylsilane (an AB_2-type monomer) via catalytic hydrosilylation reaction [95]. In a later work, Wurm et al. developed a double hydrophilic block copolymer of linear polyethylene oxide (PEO) and hbPG via anionic polymerization of ethylene oxide and subsequently

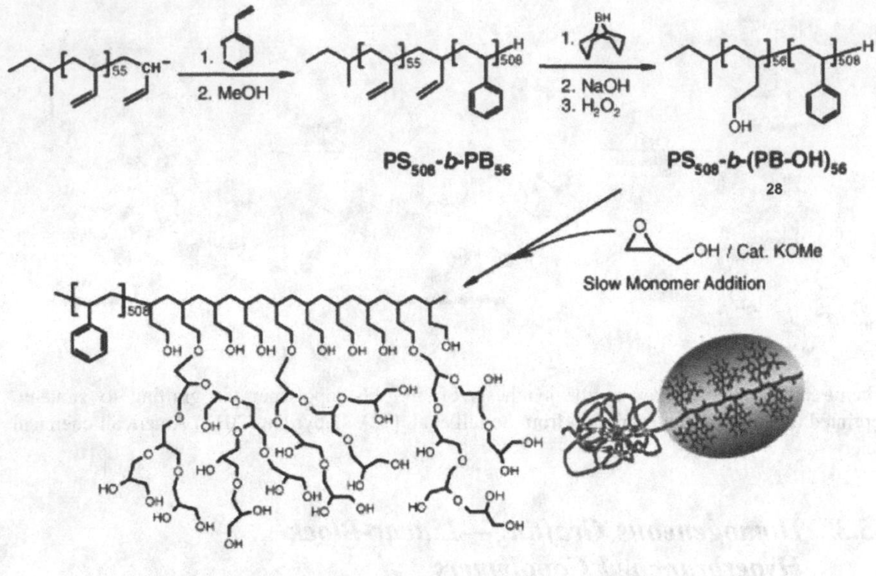

PS$_{508}$-b-PB$_{56}$

PS$_{508}$-b-(PB-OH)$_{56}$
28

△—OH / Cat. KOMe

Slow Monomer Addition

PS$_{508}$-b-(PB$_{56}$-hg-PG$_x$) 29

Scheme 3.30 Scheme showing the synthesis of a lin-b-hb copolymer via "chain first" technique. Reprinted (adapted) with permission from Barriau et al. [94]. Copyright (2005) Wiley online library

with ethoxyethyl glycidyl ether [96]. In another work, Wurm et al. synthesized amphiphilic block copolymer of linear PEO and hb PCS in a three-step process by combining anionic polymerization of allyl glycidyl ether onto PEO-OH and subsequent catalytic hydrosilylation polyaddition of an AB$_2$-type carbosilane monomer [97]. One of the disadvantages of "chain first" technique is that a certain fraction of hb homopolymer may be generated which cannot be attached to the linear blocks. However, extensive purification by precipitation may help in this regard. The "coupling strategy" is rarely employed in the synthesis of lin-b-hb copolymer because of tedious purification steps and uncontrolled coupling reactions. Limited availability of the hb dendrons with a single focal functionality reduces the probability of coupling and thus makes the synthesis procedure challenging. Yet Tao from Yan's group prepared an amphiphilic lin-b-hb copolymer through non-covalent and host–guest coupling interactions of adamantine functionalized, long alkyl chained hbPGs from ß cyclodextrin [98]. These lin-b-hb copolymers could self-assemble into unimolecular vesicles and disassembled upon the addition of a competitive host for ß-cyclodextrin. Similarly, the "core first" technique is also an unexplored strategy. However, Nuhn et al. explored the "core first" strategy to develop lin-b-hb copolymers via ROMBP of glycidol and subsequent RAFT polymerization of the hb macro-RAFT agents of functional thermoresponsive methacrylate or biocompatible methacrylamide monomers [99].

3.5.4 Heterogeneous Grafting

Often functional polymers are attached to the inorganic substrates (2D planar surfaces or spherical particles) to develop smart materials. There are four different heterogeneous hypergrafting techniques—(1) step-by-step approach, (2) graft-on-graft approach, (3) grafting from approach, and (4) grafting to approach [100]. In the step-by-step approach, which is analogous to the divergent synthesis of dendrimers, oligomeric AB_m-type branching building blocks are attached to the functionalized substrates in multiple steps. Tsubokawa et al. attached dendritic poly (amido amine); PAMAM to amino functionalized silica surfaces via an iterative alternating Michael addition of MMA to amines and amidation of the resulting esters with ethylenediamine [101]. In a similar way, Tsubokawa et al. attached PAMAM to chitosan [102]. Following their work, functional methacrylates were attached to the silica templates [103] or PAMAM to carbon blacks [104]. Unfortunately, in a step-by-step approach, only limited generations of polymers can be grown. On the contrary, in the graft-on-graft approach, as macromolecular building blocks are attached to the substrates functionalized with macromolecular groups in multiple steps, thick layers of polymers may be grown. Zhou et al. designed hb poly (acrylic acid)s; hb PAA on self-assembled organomercaptan monolayers (which is basically functionalized by reaction with α,ω-diamino-terminated poly (tert-butyl acrylate) [105]. Owing to the growing demands for one-pot synthesis strategies, grafting to and grafting from approaches have gained much attention. In the grafting to/grafting from approaches, a pre-formed hb polymer is attached to/from the substrate in one step, respectively, either through a single focal point or multiple end groups on the hb polymers. Mikhaylova et al. attached hydroxyl and carboxyl terminated hb aromatic polyesters to epoxy terminated silicon wafers via grafting to strategy [106]. In a similar fashion, Sidorenko et al. attached epoxy terminated hb polyesters to a silicon oxide substrate [107]. Paez et al. extended the grafting to strategy to gold surfaces [108]. In this novel work, in the initial step, they synthesized hbPGs with disulfide groups and different loadings of amino groups. Then, they attached these functionalized hbPGs to gold surfaces. Gold surface grafted hbPGs are potential protein resistant materials [109] and the presence of amino groups facilitated exclusive cell targeting [108].

In our lab, we are trying to develop another variation in the grafting technique by growing hb polymer chains from the surface of different substrates like polysaccharides, proteins, hydrophilic polymers, and inorganic matrices through the "Strathclyde methodology" and condensation techniques. These works are in the budding stage and thus still under consideration. We hope in future, these techniques will open up new areas of research for the design of a variety of architectural polymers.

3.5.5 Hypergrafting onto Living Cells

Of the various substrates, these days, living cells are considered as potential substrates for the attachment of hb polymers. Earlier adhesion of the hydrophilic macromolecules to cell surfaces posed serious difficulties due to repulsion between the hydrophilic components [110]. To overcome this difficulty, Rossi et al. utilized cell compatible, nonreactive additive polymers like dextran, hbPG, primary amine reactive succinimidyl succinate functionalized PEG, etc., to modify surfaces of the cells with high grafting efficiency [111]. These polymers exhibit improved penetration into the glycocalyx of the cell membranes. They carried the grafting of hbPGs on four different types of cells- RBCs, WBCs, platelets, and Jurkat cells. Interestingly, hbPG-g-live cells exhibited minimal accumulation in the organs (except liver and spleen) and reduced degree of cell membrane deformation with significant higher levels of CD47 self-protein markers which enhance their in vivo survival [112, 113]. In an attempt to improve targeting efficiency of the therapeutic stem cells to the target tissues (via intravascular injection), Jeong et al. synthesized bioactive hbPG associated stem cells [114]. They modified hbPGs with octyl chains and vasculature binding peptides (VBPs). Then they allowed modified hbPGs to bind with the cell membranes of stem cells by hydrophobic interactions and used VBPs to selectively target inflammatory endothelium or other desired tissues. Owing to the attractive features of the hb polymers and improvements in the grafting technologies, various hb polymers along with drugs/genes may be conjugated to different cells and thus may be used to improve therapies. In fact, this concept may be utilized in personalized therapies.

3.6 Conclusion

This chapter has covered some of the notable works on advanced polymerization techniques to develop hb polymers with controlled architectures and properties. The journey from FRP to SCVP, living polymerization, hypergrafting, and combination of various polymerization techniques has been discussed in lengths. Each technique has its own merits and demerits, however, as discussed; these were being selected for synthesis of explicit architectures for specific biomedical applications.

References

1. Liu J, Wang Y, Fu Q, Zhu X, Shi W (2008) Branched polymer via free radical polymerization of chain transfer monomer: a theoretical and experimental investigation. J Polym Sci A Polym Chem 46(4):1449–1459
2. O'Brien N, McKee A, Sherrington DC, Slark AT, Titterton A (2000) Facile, versatile and cost effective route to branched vinyl polymers. Polymer 41(15):6027–6031

3. Graham S, Cormack PAG, Sherrington DC (2005) One-pot synthesis of branched poly (methacrylic acid)s and suppression of the rheological "Polyelectrolyte Effect". Macromolecules 38(1):86–90

4. Liu Y, Haley JC, Deng K, Lau W, Winnik MA (2008) Synthesis of branched poly(butyl methacrylate) via semicontinuous emulsion polymerization. Macromolecules 41(12):4220–4225

5. Baudry R, Sherrington DC (2006) Synthesis of highly branched poly(methyl methacrylate)s using the "Strathclyde Methodology" in aqueous emulsion. Macromolecules 39(4):1455–1460

6. Das T, Sengupta S, Ghorai UK, Dey A, Bandyopadhyay A (2015) Sequential amphiphilic and pH responsive hyperbranched copolymer: influence of hyper branching/pendant groups on reversible self assembling from polymersomes to aggregates and usefulness in waste water treatment. RSC Adv 5(124):102932–102941

7. Dong ZM, Liu XH, Lin Y, Li YS (2008) Branched polystyrene with abundant pendant vinyl functional groups from asymmetric divinyl monomer. J Polym Sci, Part A: Polym Chem 46 (18):6023–6034

8. Dong Zm, Liu Xh, Tang Xl, Li Ys (2009) Synthesis of hyperbranched polymers with pendent norbornene functionalities via RAFT polymerization of a Novel asymmetrical divinyl monomer. Macromolecules 42, (13):4596–4603

9. Zhao T, Zhang H, Zhou D, Gao Y, Dong Y, Greiser U, Tai H, Wang W (2015) Water soluble hyperbranched polymers from controlled radical homopolymerization of PEG diacrylate. RSC Adv 5(43):33823–33830

10. Guan Z (2002) Control of polymer topology through transition-metal catalysis: synthesis of hyperbranched polymers by cobalt-mediated free radical polymerization. J Am Chem Soc 124(20):5616–5617

11. Smeets NMB (2013) Amphiphilic hyperbranched polymers from the copolymerization of a vinyl and divinyl monomer: the potential of catalytic chain transfer polymerization. Eur Polym J 49(9):2528–2544

12. Smeets NMB, Freeman MW, McKenna TFL (2011) Polymer architecture control in emulsion polymerization via catalytic chain transfer polymerization. Macromolecules 44 (17):6701–6710

13. Sato T, Sato N, Seno M, Hirano T (2003) Initiator-fragment incorporation radical polymerization of divinylbenzene in the presence of glyoxylic oxime ether: formation of soluble hyperbranched polymer. J Polym Sci A Polym Chem 41(19):3038–3047

14. Sato T, Ihara H, Hirano T, Seno M (2004) Formation of soluble hyperbranched polymer through the initiator-fragment incorporation radical copolymerization of ethylene glycol dimethacrylate with N-methylmethacrylamide. Polymer 45(22):7491–7498

15. Sato T, Arima Y, Seno M, Hirano T (2005) Initiator-fragment incorporation radical polymerization of divinyl adipate with dimethyl 2,2′-Azobis(isobutyrate): kinetics and formation of soluble hyperbranched polymer. Macromolecules 38(5):1627–1632

16. Sato T, Nomura K, Hirano T, Seno M (2006) Initiator-fragment incorporation radical polymerization of diallyl phthalate: kinetics, formation of hyperbranched polymer, and iridescent porous film thereof. J Appl Polym Sci 102(1):408–415

17. Sato T, Hashimoto M, Seno M, Hirano T (2004) Soluble hyperbranched polymer through initiator-fragment incorporation radical copolymerization of ethylene glycol dimethacrylate and α-ethyl β-N-(α-methylbenzyl) itaconamate in benzene. Eur Polym J 40(2):273–282

18. Tai H, Zheng Y, Wang W (2011) Hyperbranched copolymers synthesized by cocondensation and radical copolymerization. In: Hyperbranched polymers. Wiley, New York, pp 203–226

19. Chang HT, Frechet JMJ (1999) Proton-transfer polymerization: a new approach to hyperbranched polymers. J Am Chem Soc 121(10):2313–2314

20. Gong C, Frechet JMJ (2000) Proton transfer polymerization in the preparation of hyperbranched polyesters with epoxide chain-ends and internal hydroxyl functionalities. Macromolecules 33(14):4997–4999

21. Chen H, Jia Z, Yan D, Zhu X (2007) Thermo-responsive highly branched polyethers by proton-transfer polymerization of 1,2,7,8-diepoxyoctane and multiols. Macromol Chem Phys 208(15):1637–1645
22. Gil ES, Hudson SM (2004) Stimuli-responsive polymers and their bioconjugates. Prog Polym Sci Jpn 29(12):1173–1222
23. Gadwal I, Binder S, Stuparu MC, Khan A (2014) Dual-reactive hyperbranched polymer synthesis through proton transfer polymerization of thiol and epoxide groups. Macromolecules 47(15):5070–5080
24. Emrick T, Chang HT, Frechet JMJ (1999) An A2+B3 approach to hyperbranched aliphatic polyethers containing chain end epoxy substituents. Macromolecules 32(19):6380–6382
25. Ma Lj, Wang Hq, He Lf, Li Xy (2016) Hyperbranched epoxy resins prepared by proton transfer polymerization from an A2+B3 system. Chin J Polym Sci 29(3):300–307
26. Zhao T (2015) Controlled/living radical polymerization of multi-vinyl monomer towards hyperbranched polymers for biomedical applications (Thesis)
27. Frechet JMJ, Henmi M, Gitsov I, Aoshima S (1995) Self-condensing vinyl polymerization: an approach to dendritic materials. Science 269(5227):1080
28. Yan D, Muller AHE, Matyjaszewski K (1997) Molecular parameters of hyperbranched polymers made by self-condensing vinyl polymerization. 2. Degree of branching. Macromolecules 30(23):7024–7033
29. Paulo C, Puskas JE (2001) Synthesis of hyperbranched polyisobutylenes by inimer-type living polymerization. 1. Investigation of the effect of reaction conditions. Macromolecules 34(4):734–739
30. Knauss DM, Al-Muallem HA (2000) Polystyrene with dendritic branching by convergent living anionic polymerization. II. Approach using vinylbenzyl chloride. J Polym Sci A Polym Chem 38(23):4289–4298
31. Baskaran D (2001) Synthesis of hyperbranched polymers by anionic self-condensing vinyl polymerization. Macromol Chem Phys 202(9):1569–1575
32. Mishra M, Kobayashi S (1999) Star and hyperbranched polymers, vol. 53. CRC Press
33. Simon PFW, Radke W, Müller AHE (1997) Hyperbranched methacrylates by self-condensing group transfer polymerization. Macromol Rapid Commun 18(9):865–873
34. Chen Y, Fuchise K, Satoh T, Kakuchi T (2015) Group transfer polymerization of acrylic monomers. In: Hadjichristidis N, Hirao A (eds) Anionic polymerization: principles, practice, strength, consequences and applications. Tokyo, Springer Japan, pp 451–494
35. Simon PFW, Muller AHE (2001) Synthesis of hyperbranched and highly branched methacrylates by self-condensing group transfer copolymerization. Macromolecules 34 (18):6206–6213
36. Otsu T (2000) Iniferter concept and living radical polymerization. J Polym Sci A Polym Chem 38(12):2121–2136
37. Gigmes D (2015) Nitroxide mediated polymerization: from fundamentals to applications in materials science. R Soc Chem
38. Moad G, Rizzardo E, Solomon DH (1982) Selectivity of the reaction of free radicals with styrene. Macromolecules 15(3):909–914
39. Grubbs RB (2011) Nitroxide-mediated radical polymerization: limitations and versatility. Polym Rev 51(2):104–137
40. Hawker CJ, Frechet JMJ, Grubbs RB, Dao J (1995) Preparation of hyperbranched and star polymers by a "Living", self-condensing free radical polymerization. J Am Chem Soc 117 (43):10763–10764
41. Wang X, Gao H (2017) Recent progress on hyperbranched polymers synthesized via radical-based self-condensing vinyl polymerization. Polymers 9(6):188
42. Khan A, Malkoch M, Montague MF, Hawker CJ (2008) Synthesis and characterization of hyperbranched polymers with increased chemical versatility for imprint lithographic resists. J Polym Sci A Polym Chem 46(18):6238–6254

43. Matyjaszewski K, Gaynor SG, Greszta D, Mardare D, Shigemoto T, Wang J-S (1995) Unimolecular and bimolecular exchange reactions in controlled radical polymerization. Macromol Symp 95(1):217–231
44. Patten TE, Matyjaszewski K (1998) Atom transfer radical polymerization and the synthesis of polymeric materials. Adv Mater 10(12):901–915
45. Matyjaszewski K, Tsarevsky NV (2014) Macromolecular engineering by atom transfer radical polymerization. J Am Chem Soc 136(18):6513–6533
46. Gao H, Matyjaszewski K (2009) Synthesis of functional polymers with controlled architecture by CRP of monomers in the presence of cross-linkers: from stars to gels. Prog Polym Sci 34(4):317–350
47. Gaynor SG, Edelman S, Matyjaszewski K (1996) Synthesis of branched and hyperbranched polystyrenes. Macromolecules 29(3):1079–1081
48. Graff RW, Wang X, Gao H (2015) Exploring self-condensing vinyl polymerization of inimers in microemulsion to regulate the structures of hyperbranched polymers. Macromolecules 48(7):2118–2126
49. Matyjaszewski K, Gaynor SG, Kulfan A, Podwika M (1997) Preparation of hyperbranched polyacrylates by atom transfer radical polymerization. 1. Acrylic AB* monomers in "Living" radical polymerizations. Macromolecules 30(17):5192–5194
50. Muthukrishnan S, Jutz G, Andre X, Mori H, Muller AHE (2005) Synthesis of hyperbranched glycopolymers via self-condensing atom transfer radical copolymerization of a sugar-carrying acrylate. Macromolecules 38(1):9–18
51. Amin A, El-Gaffar MA (2008) Synthesis of novel polyamide-hyperbranched polymers via self condensing atom transfer radical polymerization. Polym Plast Technol 47(10):984–990
52. Tsarevsky NV, Huang J, Matyjaszewski K (2009) Synthesis of hyperbranched degradable polymers by atom transfer radical (Co)polymerization of inimers with ester or disulfide groups. J Polym Sci A Polym Chem 47(24):6839–6851
53. Jakubowski W, Matyjaszewski K (2006) Activators regenerated by electron transfer for atom-transfer radical polymerization of (Meth)acrylates and related block copolymers. Angew Chem Int Ed 45(27):4482–4486
54. Elsen AM, Burdynska J, Park S, Matyjaszewski K (2013) Activators regenerated by electron transfer atom transfer radical polymerization in miniemulsion with 50 ppm of copper catalyst. ACS Macro Lett 2(9):822–825
55. Boyer C, Corrigan NA, Jung K, Nguyen D, Nguyen TK, Adnan NNM, Oliver S, Shanmugam S, Yeow J (2016) Copper-mediated living radical polymerization (atom transfer radical polymerization and copper(0) mediated polymerization): from fundamentals to bioapplications. Chem Rev 116(4):1803–1949
56. Matyjaszewski K, Pyun J, Gaynor SG (1998) Preparation of hyperbranched polyacrylates by atom transfer radical polymerization, 4. The use of zero-valent copper. Macromol Rapid Commun 19(12):665–670
57. Konkolewicz D, Wang Y, Krys P, Zhong M, Isse AA, Gennaro A, Matyjaszewski K (2014) SARA ATRP or SET-LRP. End of controversy? Polym Chem 5(15):4396–4417
58. Xue X, Li F, Huang W, Yang H, Jiang B, Zheng Y, Zhang D, Fang J, Kong L, Zhai G, Chen J (2014) Quadrangular prism: a unique self-assembly from amphiphilic hyperbranched PMA-b-PAA. Macromol Rapid Commun 35(3):330–336
59. Moad G, Rizzardo E, Thang SH (2006) Living radical polymerization by the RAFT—a first update. Aust J Chem 59(10):669–692
60. McLeary JB, Calitz FM, McKenzie JM, Tonge MP, Sanderson RD, Klumperman B (2005) A 1H NMR investigation of reversible addition-fragmentation chain transfer polymerization kinetics and mechanisms. Initialization with different initiating and leaving groups. Macromolecules 38(8):3151–3161
61. Alfurhood JA, Bachler PR, Sumerlin BS (2016) Hyperbranched polymers via RAFT self-condensing vinyl polymerization. Polym Chem 7(20):3361–3369
62. Wang Z, He J, Tao Y, Yang L, Jiang H, Yang Y (2003) Controlled chain branching by RAFT-based radical polymerization. Macromolecules 36(20):7446–7452

63. Rikkou-Kalourkoti M, Elladiou M, Patrickios CS (2015) Synthesis and characterization of hyperbranched amphiphilic block copolymers prepared via self-condensing RAFT polymerization. J Polym Sci Part A Polym Chem 53(11):1310–1319

64. Heidenreich AJ, Puskas JE (2008) Synthesis of arborescent (Dendritic) polystyrenes via controlled inimer-type reversible addition-fragmentation chain transfer polymerization. J Polym Sci A Polym Chem 46(23):7621–7627

65. Carter S, Rimmer S, Sturdy A, Webb M (2005) Highly branched stimuli responsive poly [(N-isopropyl acrylamide)-co-(1,2-propandiol-3-methacrylate)]s with protein binding functionality. Macromol Biosci 5(5):373–378

66. Ghosh Roy S, De P (2014) Facile RAFT synthesis of side-chain amino acids containing pH-responsive hyperbranched and star architectures. Polym Chem 5(21):6365–6378

67. Han J, Li S, Tang A, Gao C (2012) Water-soluble and clickable segmented hyperbranched polymers for multifunctionalization and novel architecture construction. Macromolecules 45(12):4966–4977

68. Bai Lb, Zhao K, Wu Yg, Li Wl, Wang Sj, Wang Hj, Ba Xw, Zhao Hc (2014) A new method for synthesizing hyperbranched polymers with reductive groups using redox/RAFT/SCVP. Chin J Polym Sci 32(4):385–394

69. Delduc P, Tailhan C, Zard SZ (1988) A convenient source of alkyl and acyl radicals. J Chem Soc Chem Commun 4:308–310

70. Zhou X, Zhu J, Xing M, Zhang Z, Cheng Z, Zhou N, Zhu X (2011) Synthesis and characters of hyperbranched poly (vinyl acetate) by RAFT polymeraztion. Eur Polym J 47(10):1912–1922

71. Perrier S, Takolpuckdee P (2005) Macromolecular design via reversible addition–fragmentation chain transfer (RAFT)/xanthates (MADIX) polymerization. J Polym Sci A Polym Chem 43(22):5347–5393

72. Moad G (2006) The emergence of RAFT polymerization. Aust J Chem 59(10):661–662

73. Postma A, Davis TP, Moad G, O'Shea MS (2005) Thermolysis of RAFT-synthesized polymers. A convenient method for trithiocarbonate group elimination. Macromolecules 38(13):5371–5374

74. Gao C, Yan D (2004) Hyperbranched polymers: from synthesis to applications. Prog Polym Sci 29(3):183–275

75. Wilms D, Nieberle J, Frey H (2011) Ring-opening multibranching polymerization. In: Hyperbranched Polymers. Wiley, New York, pp 175–202

76. Hauser M (1969) Alkylene imines. In: Frisch KC, Reegen SL, Dekker M (eds) Ring-opening polymerization. New York

77. Vandenberg EJ (1985) Polymerization of glycidol and its derivatives: a new rearrangement polymerization. J Polym Sci Polym Chem Ed 23(4):915–949

78. Sunder A, Hanselmann R, Frey H, Mulhaupt R (1999) Controlled synthesis of hyperbranched polyglycerols by ring-opening multibranching polymerization. Macromolecules 32(13):4240–4246

79. Goodwin A, Baskaran D (2012) Inimer mediated synthesis of hyperbranched polyglycerol via self-condensing ring-opening polymerization. Macromolecules 45(24):9657–9665

80. Rokicki G, Rakoczy P, Parzuchowski P, Sobiecki M (2005) Hyperbranched aliphatic polyethers obtained from environmentally benign monomer: glycerol carbonate. Green Chem 7(7):529–539

81. Kainthan RK, Janzen J, Kizhakkedathu JN, Devine DV, Brooks DE (2008) Hydrophobically derivatized hyperbranched polyglycerol as a human serum albumin substitute. Biomaterials 29(11):1693–1704

82. Gao X, Zhang X, Zhang X, Wang Y, Sun L, Li C (2011) Amphiphilic polylactic acid-hyperbranched polyglycerol nanoparticles as a controlled release system for poorly water-soluble drugs: physicochemical characterization. J Pharm Pharmacol 63(6):757–764

83. Gao S, Guan Q, Chafeeva I, Brooks DE, Nguan CYC, Kizhakkedathu JN, Du C (2015) Hyperbranched polyglycerol as a colloid in cold organ preservation solutions. PloS One 10(2):e0116595

84. Wilms D, Stiriba SE, Frey H (2010) Hyperbranched polyglycerols: from the controlled synthesis of biocompatible polyether polyols to multipurpose applications. Acc Chem Res 43(1):129–141

85. Liu J, Wu X, Liu Y, Xu Y, Huang Y, Xing C, Wang X (2016) High-glucose-based peritoneal dialysis solution induces the upregulation of VEGF expression in human peritoneal mesothelial cells: the role of pleiotrophin. Int J Mol Med 32

86. Schull C, Frey H (2013) Grafting of hyperbranched polymers: from unusual complex polymer topologies to multivalent surface functionalization. Polymer 54(21):5443–5455

87. Bergbreiter DE, Kippenberger AM (2006) Hyperbranched surface graft polymerizations. In: Jordan R (ed) Surface-Initiated Polymerization II. Springer, Berlin Heidelberg: Berlin, Heidelberg, pp 1–49

88. Popeney CS, Lukowiak MC, Bottcher C, Schade B, Welker P, Mangoldt D, Gunkel G, Guan Z, Haag R (2012) Tandem coordination, ring-opening, hyperbranched polymerization for the synthesis of water-soluble core-shell unimolecular transporters. ACS Macro Lett 1 (5):564–567

89. Xu Y, Gao C, Kong H, Yan D, Luo P, Li W, Mai Y (2004) One-pot synthesis of amphiphilic core-shell suprabranched macromolecules. Macromolecules 37(17):6264–6267

90. Kuo PL, Ghosh SK, Liang WJ, Hsieh YT (2001) Hyperbranched polyethyleneimine architecture onto poly(allylamine) by simple synthetic approach and the chelating characters. J Polym Sci A Polym Chem 39(17):3018–3023

91. Schull C, Frey H (2012) Controlled synthesis of linear polymers with highly branched side chains by "Hypergrafting": poly(4-hydroxy styrene)-graft-hyperbranched polyglycerol. ACS Macro Lett 1(4):461–464

92. Schull C, Nuhn L, Mangold C, Christ E, Zentel R, Frey H (2012) Linear-hyperbranched graft-copolymers via grafting-to strategy based on hyperbranched dendron analogues and reactive ester polymers. Macromolecules 45(15):5901–5910

93. Wurm F, Frey H (2011) Linear-dendritic block copolymers: the state of the art and exciting perspectives. Prog Polym Sci 36(1):1–52

94. Barriau E, García Marcos A, Kautz H, Frey H (2005) Linear-hyperbranched amphiphilic AB Diblock copolymers based on polystyrene and hyperbranched polyglycerol. Macromol Rapid Commun 26(11):862–867

95. Marcos AGa, Pusel TM, Thomann R, Pakula T, Okrasa L, Geppert S, Gronski W, Frey H (2006) Linear-hyperbranched block copolymers consisting of polystyrene and dendritic poly (carbosilane) block. Macromolecules 39(3):971–977

96. Wurm F, Nieberle Jr, Frey H (2008) Double-hydrophilic linear-hyperbranched block copolymers based on poly(ethylene oxide) and poly(glycerol). Macromolecules 41(4):1184–1188

97. Wurm F, Schule H, Frey H (2008) Amphiphilic linear-hyperbranched block copolymers with linear poly(ethylene oxide) and hyperbranched poly(carbosilane) block. Macromolecules 41(24):9602–9611

98. Tao W, Liu Y, Jiang B, Yu S, Huang W, Zhou Y, Yan D (2012) A linear-hyperbranched supramolecular amphiphile and its self-assembly into vesicles with great ductility. J Am Chem Soc 134(2):762–764

99. Nuhn L, Schull C, Frey H, Zentel R (2013) Combining ring-opening multibranching and RAFT polymerization: multifunctional linear-hyperbranched block copolymers via hyperbranched macro-chain-transfer agents. Macromolecules 46(8):2892–2904

100. Peleshanko S, Tsukruk VV (2008) The architectures and surface behavior of highly branched molecules. Prog Polym Sci 33(5):523–580

101. Tsubokawa N, Ichioka H, Satoh T, Hayashi S, Fujiki K (1998) Grafting of 'dendrimer-like' highly branched polymer onto ultrafine silica surface. React Funct Polym 37(1):75–82

102. Tsubokawa N, Takayama T (2000) Surface modification of Chitosan powder by grafting of 'dendrimer-like' hyperbranched polymer onto the surface. React Funct Polym 43(3):341–350

103. Wang G, Fang Y, Kim P, Hayek A, Weatherspoon MR, Perry JW, Sandhage KH, Marder SR, Jones SC (2009) Layer-by-layer dendritic growth of hyperbranched thin films for surface sol-gel syntheses of conformal, functional, nanocrystalline oxide coatings on complex 3D (Bio)silica templates. Adv Funct Mater 19(17):2768–2776

104. Tsubokawa N, Satoh T, Murota M, Sato S, Shimizu H (2001) Grafting of hyperbranched poly(amidoamine) onto carbon black surfaces using dendrimer synthesis methodology. Polym Adv Tech 12(10):596–602

105. Zhou Y, Bruening ML, Bergbreiter DE, Crooks RM, Wells M (1996) Preparation of hyperbranched polymer films grafted on self-assembled monolayers. J Am Chem Soc 118(15):3773–3774

106. Mikhaylova Y, Pigorsch E, Grundke K, Eichhorn KJ, Voit B (2004) Surface properties and swelling behaviour of hyperbranched polyester films in aqueous media. Macromol Symp 210(1):271–280

107. Sidorenko A, Zhai XW, Greco A, Tsukruk VV (2002) Hyperbranched polymer layers as multifunctional interfaces. Langmuir 18(9):3408–3412

108. Paez JI, Brunetti V, Strumia MC, Becherer T, Solomun T, Miguel J, Hermanns CF, Calderon M, Haag R (2012) Dendritic polyglycerolamine as a functional antifouling coating of gold surfaces. J Mater Chem 22(37):19488–19497

109. Siegers C, Biesalski M, Haag R (2004) Self-assembled monolayers of dendritic polyglycerol derivatives on gold that resist the adsorption of proteins. Chem Eur J 10(11):2831–2838

110. Dhalluin C, Ross A, Leuthold LA, Foser S, Gsell B, Muller F, Senn H (2005) Structural and biophysical characterization of the 40 kDa PEGâˆ'Interferon-Î ± 2a and its individual positional isomers. Bioconjug Chem 16(3):504–517

111. Rossi NAA, Constantinescu I, Brooks DE, Scott MD, Kizhakkedathu JN (2010) Enhanced cell surface polymer grafting in concentrated and nonreactive aqueous polymer solutions. J Am Chem Soc 132(10):3423–3430

112. Chapanian R, Constantinescu I, Brooks DE, Scott MD, Kizhakkedathu JN (2012) In vivo circulation, clearance, and biodistribution of polyglycerol grafted functional red blood cells. Biomaterials 33(10):3047–3057

113. Chapanian R, Constantinescu I, Rossi NAA, Medvedev N, Brooks DE, Scott MD, Kizhakkedathu JN (2012) Influence of polymer architecture on antigens camouflage, CD47 protection and complement mediated lysis of surface grafted red blood cells. Biomaterials 33 (31):7871–7883

114. Jeong JH, Schmidt JJ, Kohman RE, Zill AT, DeVolder RJ, Smith CE, Lai M-H, Shkumatov A, Jensen TW, Schook LG, Zimmerman SC, Kong H (2013) Leukocyte-mimicking stem cell delivery via in situ coating of cells with a bioactive hyperbranched polyglycerol. J Am Chem Soc 135(24):8770–8773

Chapter 4
Structure–Property Relationship of Hyperbranched Polymers

4.1 Introduction to Intrinsic Properties of Hyperbranched Polymers

The field of hyperbranched polymers has been explored widely and has been a great topic for researchers in the last two decades because these compounds possess new, remarkable characteristics that strongly influence material properties and have opened new application fields [1]. Therefore, simultaneously along with the synthesis of various kinds of hyperbranched polymers from different combinations of monomers, special emphasis has been taken to establish its structure-property relationship by proper characterization and understanding of the structure of the hyperbranched polymers and its effect on its physical and chemical properties. With the wide spread of hyperbranched polymers into the global market, the complex branched hyperbranched polymer structures have also presented an enormous challenge for full structure characterization and have shown limitations of current techniques.

There are several reviews done on the different kinds of hyperbranched polymers, from AB_x type monomer combination to A_2+B_y type monomer pairs, SCVP, and in general all the possible highly branched polymers. All these studies reveal that hyperbranched polymers are very complex structures with multidimensional, broad distributions, e.g., in molar mass, degree of branching (DB), or chemical structure. Hyperbranched polymers, as already known to us, have unique and special structural properties differing from its linear analog. Polymeric chains of hyperbranched polymers diverge unsymmetrically in three-dimensional space from a point and they are shaped as a branching tree [2]. Molecular weight is not sufficient to characterize these polymers, as the possibility of various structures increases with increasing degree of polymerization [3]. Thus, they have different influence of their structure on the properties which will be discussed further in details. An overview of the different methods for the structural, thermal, solution, and bulk characterization of the hyperbranched polymers are given in this chapter.

© Springer Nature Singapore Pte Ltd. 2018
A. Bandyopadhyay et al., *Hyperbranched Polymers for Biomedical Applications*,
Springer Series on Polymer and Composite Materials,
https://doi.org/10.1007/978-981-10-6514-9_4

4.2   Degree of Branching

The DB is an important parameter to characterize hyperbranched polymers. It measures the extent as well as the perfectness of the branches [4]. Initially, Fréchet and Hawker defined it simply as $(1 - a_2)$, where a_2 was the fraction of linear segments in the system [5]. At present, DB is the most commonly used parameter to quantify the architecture of highly branched polymers. DB is different from the term "branching coefficient" where the latter defines the probability of a link in a branch unit to lead to another branch unit [6]. Later Kim and Webster estimated the extent of branching by simply giving the ratio between the molar fraction of fully branched monomers and that of all possible branching sites [7], where as Hölter et al. published a great number of research articles based on the different methods to estimate and control DB. Another parameter, "average number of branches," was coined by Frey's group which defines the number of branches emerging from units divided by the total number of units [8].

4.2.1   Determination of DB

Hyperbranched polymers consist of three different types of repeating units like dendritic unit, linear unit and the terminal unit depending on the number of unreacted functional groups present unlike the dendrimers where there is only dendritic and terminal units. In 1991 Fréchet described DB to analyze the structure of hyperbranched polymers as per the given equation

$$\mathrm{DB} = \frac{D+T}{D+T+L} \tag{4.1}$$

where D is the number of dendritic units, T is the number of terminal units and L is the number of linear units, respectively (represented in Fig. 4.1).

DB is an important characteristic to differentiate between hyperbranched polymers and dendrimers as for dendrimers, due to its strictly symmetrical structure, has DB nearly 1 where as for hyperbranched polymers the DB < 1 [9]. It is concluded that hyperbranched polymers have DB parameter = 0.4–0.8 which distinguishes its properties from conventional linear and cross-linked polymers. The characteristic properties of hyperbranched polymer are high solubility and thermodynamic compatibility, low viscosity of solutions, absence of entanglements and ability to act as nanocarriers which are all dependent on the DB in the corresponding hyperbranched polymer.

By using statistical analysis, Holter and his coworkers tried to find a correlation between the DB and the monomer conversion [10]. They tried to estimate the maximum value of DB that can be achieved by changing the f for AB_f monomers. It was observed that the DB value is dependent linearly with the conversion and it is

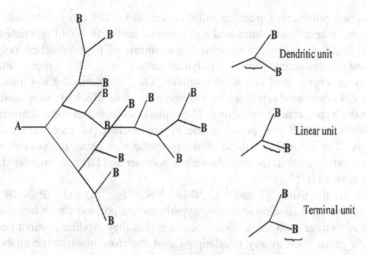

Fig. 4.1 Different segment types present in hyperbranched polymers

not dependent on the functionality of the AB_f monomer [11]. Even in self-condensing vinyl polymerization, the value of DB was found quite low which was not improved by using multifunctional initiator in the system.

There have been several attempts to increase this DB in the hyperbranched polymers, such as: (1) further polymerization of dendrons containing preformed dendritic units or postpolymerization modification of the polymers, (2) polymerization of AB_x type of monomers initiating from a multifunctional core, and (3) selection of those monomers whose reactivity increases with conversion. But the most efficient way to increase the DB for a hyperbranched polymer system is the slow addition of monomers which is effective in case of both condensation and self-condensing polymerization technique [9].

There are different methods to determine the DB in a hyperbranched polymer directly or indirectly, as it is very important to characterize the architecture of the polymer. NMR spectroscopy is the strongest tool to determine DB. Apart from the conventional ^{1}HNMR, ^{13}C, ^{15}N, and other NMR spectroscopy, there are also other methods to determine DB such as viscometry and by quantitative analysis of degraded products if the polymer chain contains degradable linkages [12, 13].

4.2.2   Methods to Determine DB

4.2.2.1   Direct Method

Branching has a large influence on the characteristics of a polymer; therefore, the characterization of a polymer architecture is very important. Nuclear magnetic resonance (NMR) is a very important tool to determine the structure for hyperbranched polymers. Direct quantitative values for DB can be obtained from NMR spectra where linear, branched and terminal units can be differentiated.

Fréchet and his coworkers reported the synthesis of hyperbranched polyester using self-condensation vinyl polymerization of AB_2 monomer, as 3,5-dihydroxybenzoic acid and its derivatives. They synthesized a low molecular weight model compound representing the repeating units which are supposed to be present in the hyperbranched structure. These model compounds were characterized by ^{13}C NMR and different peaks in the hyperbranched polymer were assigned accordingly. The integrals of those peaks gave the idea about the content of the different repeating units in the hyperbranched polymer and DB was calculated using the above given Eq. (4.1).

Apart from the usual ^1H and ^{13}C NMR, ^{15}N, ^{19}F, ^{29}Si, and ^{31}P NMR spectroscopy is also used for characterizing hyperbranched polymers with heteroatoms. The main advantage with these NMR spectra is that they produce distinct peak for the specific atom without any overlapping and therefore quantitative analysis of these peaks is much less complicated [14].

Figure 4.2 gives an example of determining the structural units of hyperbranched polyetheramide [15] from NMR spectroscopy. The polymer is obtained from AB_2 type monomer. Though for A_2+B_3 pair of monomers, the determination of the structural units is not that easy and a number of possible monomer combinations are given in Scheme 4.1. Model compounds need to be established for

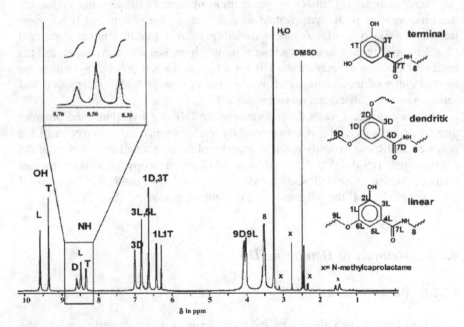

Fig. 4.2 ^1H NMR spectrum for determination of the amount of T, L, and D units in hyperbranched poly(etheramide) made by ring-opening polymerization of an AB_2 monomer. Reprinted with permission from Ref. [15]. Copyright (1999) Wiley VCH

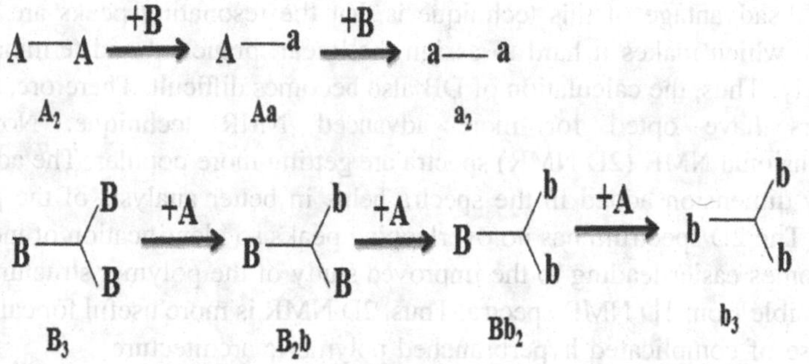

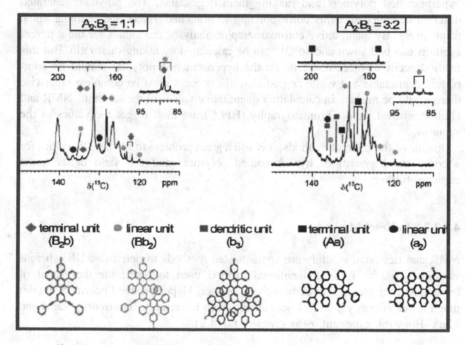

Scheme 4.1 Possible reactions of the functional groups in A_2+B_3 systems and the resulting structural units

Fig. 4.3 ^{13}C NMR spectra and different structural units of polyphenylenes produced by different ratios of A_2+B_3 monomers (compare Scheme 4.1). Reprinted with permission from Ref. [16]. Copyright (2006) Wiley VCH

proper evaluation of the structure from the spectroscopy results. Figure 4.3 demonstrates ^{13}C NMR of hyperbranched polyphenylenes having A_2+B_3 in different ratio. The spectra reveal different structural units based on the model compounds [16].

In general, the one-dimensional NMR (1D NMR) is the most acceptable and widely practiced method to quantify the structure of a hyperbranched polymer. But the main disadvantage of this technique is that the resonating peaks are not differentiable which makes it hard to assign the linear, branched and terminal units, respectively. Thus, the calculation of DB also becomes difficult. Therefore, recently researchers have opted for more advanced NMR technique. Nowadays, two-dimensional NMR (2D NMR) spectra are getting more popular. The additional frequency dimension added in the spectra helps in better analysis of the polymer structure. The 2D spectrum has no overlapping peaks so identification of individual units becomes easier leading to the improved study of the polymer structure which is not possible from 1D NMR spectra. Thus, 2D NMR is more useful for calculating DB in case of complicated hyperbranched polymeric architecture.

Another direct technique, proposed by Kambouris and Hawker [17] and Bolton and Wooley [18] is quite unique and different from the previously discussed method. They degraded the macromolecule by altering the functional groups of the hyperbranched polymers and making them degradable. The polymer generated different degraded subunits corresponding to branched, linear and terminal units, respectively. By quantitative chromatographic analysis, the values for the different subunits can be known and so DB can be calculated by taking their ratio. But this method needs to meet few criteria: the hyperbranched polymer should be completely degradable and while degradation, the process must be complete otherwise there would be mistake in calculating quantitative values of the subunits. NMR and High-Pressure Liquid Chromatography (HPLC) are used to get the values of the subunits.

Inspite of these direct methods, it is still a great problem to get the DB values for a complicated system of hyperbranched polymers and this field needs to be explored further.

4.2.2.2   Indirect Method

NMR and degradation study are sophisticated methods to determine DB whereas viscometry is the most conventional method used to determine the extent of branching indirectly in a hyperbranched polymer. Hyperbranched polymers exhibit much lower viscosity than its linear analog, with lower intrinsic viscosity value and Mark–Houwink exponent α (as given in Eq. 4.2).

$$[\eta] = KM_v^{\alpha} \tag{4.2}$$

where $[\eta]$ is the intrinsic viscosity, M_v is the viscosity average molecular weight, K and α are the Mark–Houwink constant.

The lowering of the intrinsic viscosity values in case of hyperbranched polymers is due to the more compact structure and lack of entanglements. Other than viscometry, contraction factor and radius of gyration (R_g) values for the polymer also gives an idea about the structure of the branched polymer.

Along with the aforesaid techniques, sometimes Fourier Transfer Infrared Spectroscopy (FT-IR) is also used to interpret about the architecture of the hyperbranched polymers.

Another parameter, Weiner Index is also used to quantify a hyperbranched polymer, alongside DB [19]. It is defined as the sum of the distances between all pairs of units in the molecular assembly. It is generally higher for linear polymeric chain and the value decreases with increasing branches and compactness in the molecular structure. The Weiner Index is related to the square of the Rg of polymer molecules, but only if the links counted by the index are Gaussian subchains.

DB is one of the most important parameters for hyperbranched polymers because it has a strong influence on the polymeric properties such as free volume, chain entanglement, mean-square R_g, glass transition temperature (T_g), degree of crystallization (DC), capability of encapsulation, mechanical strength, melting/solution viscosity, biocompatibility, and self-assembly behaviors. Details of the effect of DB on the other polymeric properties are discussed further in the chapter.

4.3 Influence of the Branching Architecture on the End Properties

DB, being a very important parameter for hyperbranched polymers, it greatly influences the physical and chemical properties of the hyperbranched polymers. Hyperbranched polymers have gained so much attention in the field of polymeric research due to its unique properties compared to its linear grade. It is stated that the application efficiency of hyperbranched polymers is so high because of its low molecular entanglement, low melting/solution viscosity, weak mechanical strength, high solubility, highly reactive functional groups, the excellent capacity of encapsulation for guest molecules, and unusual self-assembly behaviors. The properties that attribute to these characteristics are its three-dimensional architecture with irregularly branched topology with DB < 1.0 (normally, 0.4–0.6), high polydispersity of Mw (normally, PDI > 3.0), and the presence of a high number of functional groups linked at both the linear and terminal units [20]. Apart from DB, there are several other parameters that determine the specific properties of a hyperbranched polymer, such as type of functional groups, length, and flexibility of linear parts, and molecular mass of the characteristic polymer.

Hyperbranched polymers can be differentiated depending on its chemical properties and topology. Chemical properties are determined by the functional groups present in the structure but the changes in properties related to the structure of the hyperbranched polymers are considered more significant rather than the chemical properties. As the proper evaluation of the structure–property relationship for hyperbranched polymers is still lacking, recently researchers have put more focus on this topic to know why and to what extent the architecture affect the properties.

4.4   Solution Properties with Special Reference to Hyperbranched Architecture

Solubility has been a major problem for conventional polymers over the years. Its macromolecular entangled structure hindered any kind of solvent penetration in the structure making polymers very hard to solubilize. Researchers have tried over the years to overcome this issue by various means such as introducing polar functional groups, by controlling the chain length, etc. One of the most basic properties of hyperbranched polymers that differentiate it from those of linear analogs is the high solubility induced by the branched backbone. Kim and Webster [21] reported that hyperbranched polyphenylenes showed a much better solubility in various solvents as compared to linear polyphenylenes, which have a very poor solubility. The highly polar carboxylate groups present at the terminal end of the hyperbranched polymers made it more soluble than the linear ones, even water soluble. This high solubility makes hyperbranched polymers a great choice in biomedical applications. Water soluble hyperbranched polymers are used as vehicles for various delivery purposes in biomedicine.

The solution theory of the intrinsic viscosity of branched polymers is based on the Flory–Fox equation:

$$[\eta] = \phi \left(\frac{R_g^3}{M} \right) \qquad (4.3)$$

This relationship can easily be applied to linear polymers due to the fact that the draining factor ϕ is independent on the size of the molecules and is a constant but for the draining parameter increases with increasing segment density [22]. Therefore for hyperbranched polymers, Mark–Houwink-Sakurada equation (Eq. 4.2) is taken into consideration which showed an unusual behavior—unlike conventional linear, star or highly branched polymers, the equation is not dependent on the increasing molar mass in case of hyperbranched polymers.

Hyperbranched polymers also differ in solution behavior in comparison to the linear grades. Both solution and melt viscosities are considerably lower in case of the hyperbranched polymer. The relationship between intrinsic viscosity and molecular mass of a polymer is given by Mark–Houwink–Sakurada equation as given in Eq. 4.2. Intrinsic viscosity increases alongwith the increment of the molecular mass in case of both linear and hyperbranched polymers, but due to extensive chain entanglements, the increase in intrinsic viscosity is drastically increased in linear polymers after critical molar mass is reached. In case of hyperbranched polymers there is no critical molar mass due to lack of entanglements. Fréchet demonstrated the relationship between intrinsic viscosity and molecular weight, related to its architecture as shown in Fig. 4.4 [23].

Fig. 4.4 Generalized description of the intrinsic viscosity as function of molar mass for linear polymers, hyperbranched polymers and dendrimers as described by Fréchet

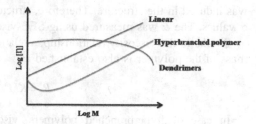

By viscosity measurements and scattering methods, information about the hydrodynamic radius and R_g can also be obtained. These results conclude about the molecular shape, density, and self-similarity of the hyperbranched polymers [24].

α has significance in characterizing hyperbranched polymers. As it is already known that k and α from Eq. 4.2 are constants specificifying a certain solvent–polymer combination at a certain temperature. Values of α can be related to chain extensions in dilute solutions. Mark–Houwink constant signifies the shape and compactness of a polymer in a solvent. Hyperbranched polymers, having a dense and compact globular structure, have a lower α range than the linear polymers. Generally the value range of $0.3 < \alpha < 0.5$ was found for hb polymers, whereas for a linear statistical coil in a good solvent, values of $0.5 < \alpha < 1$ are recorded [25].

In Fig. 4.5, the dependence of α on DB was measured for two different sets of aromatic–aliphatic polyesters. It is expected that with lowering of DB, α will increase. In first case, AB_2 type monomer was taken along with AB type which resulted in decreasing of branches. At lower DB, it showed lowering of α. The probable reason for it was may be that the stiff AB_2 monomer was taken with more flexible AB monomer which leads to a stable flexible and more compact conformation with a lower value of α. Again when AB_2 monomer was taken with monoprotected $AB_{2,mp}$ monomer, there was also lowering of DB but no flexibility

Fig. 4.5 KMHS plot of aliphatic and aliphatic–aromatic polyester with nonpolar end groups. Reprinted with permission from Ref. [26]. Copyright (2009) American Chemical Society

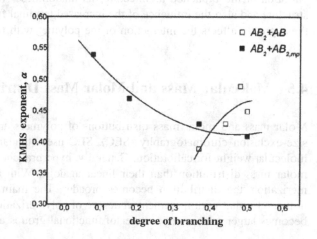

was induced in the structure. Therefore, structures were more compact with higher α values. The α was measured using SEC-viscosity detector [26].

Similar to the viscosity relationship, a relationship between the R_g and the molar mass of the polymer is also established as:

$$R_g = KM^\nu \tag{4.4}$$

In case of hyperbranched polymers, viscosity is reduced due to branched structure; similarly, R_g also decreases as HBs have dense compact structure with much lower volume. Along with that, as obtained by the contraction factors g and g' hyperbranched polymers exhibits a stronger shrinking effect in comparison to long-chain branched polymers [27]:

$$g = \frac{R^2_{g.branched}}{R^2_{g.linear}} \tag{4.5}$$

$$g' = \frac{[\eta]_{branched}}{[\eta]_{linear}} \tag{4.6}$$

The relationship between both the contraction factors have been established theoretically as well as experimentally as:

$$g' = g^b$$

This value of b varies in case of hyperbranched polymers due to varying compactness and interaction of the polymer with the solvent, and can be used to compare it with its linear analog.

Though several new and interesting relationships have been discussed in this section, still a generalized theory regarding the solution properties of the hyperbranched polymers cannot be established till date. The reason behind this is the variation of the branched architecture, its randomness, variation of the backbone structure and also the influence of the increased terminal functional groups as all of these greatly affects the interaction of the polymer with the solvent.

4.5 Molecular Mass and Molar Mass Distribution

Molar mass and molar mass distributions of polymers are generally estimated by size-exclusion chromatography (SEC). SEC uses standard polymers with known molecular weight for calibration. Generally, hyperbranched polymers show a broad molar mass distribution than their linear analogs. With a higher degree of polymerization, the distribution becomes broader. With a higher degree of polymerization, the distribution becomes broader. The main reason behind this phenomenon is that with increment in degree of polymerization, the polymer molecule becomes larger with more number of functional groups at the terminal end which reacts further to give a wider molecular weight distribution [28]. Apart from con-densation reactions, in case of self-condensing vinyl polymerisation also the molar weight distribution is strongly dependent on the degree of polymerisation [29].

The broad molar mass can also be controlled by other factors such as slow monomer addition [30] or by taking a multifunctional core molecule [31].

Again for A_2+B_3, system, the molar mass distribution gets affected by various factors: functionality conversion, the ratio A/B, the amount of the added A_2 monomer [32]. Precisely, it can be concluded that molar mass of hyperbranched polymers is higher than that of its linear analog since due to highly branched structure the molecular density is increased largely.

Size-exclusion chromatography (SEC) with differential refractive index detec-tion (DRI) or UV-detection and subsequent calibration with a linear polymer standard is the most well-known technique to determine the molecular weight and its distribution.

Light-scattering techniques also give knowledge about the weight average molecular weight of a polymer. Two main types are used: multiangle laser light scattering (MALLS) and low-angle laser light scattering (LALLS). In case of MALLS, it is less sensitive toward impurities, and information regarding molecular size and polydispersity is obtained simultaneously since the measurements are performed at several angles simultaneously and therefore it is much preferred over LALLS to get the details of the molar mass distribution.

But using static light-scattering technique alone is not sufficient to estimate the broad polydispersity of hyperbranched polymers, as light-scattering techniques are able to give the molecular weight provided there is no aggregate in the solution. But in reality, it is not possible thus it has to be used in combination with SEC. The total combination of SEC–DRI–MALLS is needed to give the overall molecular weight distribution of complexed polymeric molecule such as hyperbranched polymer.

Results obtained from light-scattering techniques are dependent on the specific refractive increment, dn/dc according to Eq. (4.7)

$$R_\theta = \frac{4\pi^2 n_0^2 \left(\frac{dn}{dc}\right)^2}{\lambda_0^4 N_L} cM \qquad (4.7)$$

For lower mass region of the polymer, high specific refractive increment is required for proper detection in MALLS. Therefore, high concentration is required in the solvent which is a big limitation for SEC analysis. Therefore, the combination is suitable for high molar mass region of the hyperbranched polymer, but unable to analyze the lower mass region. So alternative ways were thought to overcome the demerits of this method and get the complete analysis of the hyperbranched polymers [33].

The newly developed matrix-assisted laser desorption-ionization time-offlight (MALDI-TOF) technique has been used for hyperbranched polymers with good results. The technique suffers from some uncertainties due to it requires monodisperse samples for proper characterization. Therefore, polydisperse hyperbranched polymers need to be fractionated to monodisperse fractions and then measured using MALDI-TOF.

4.6 Bulk Properties

4.6.1 Thermal Properties

Thermal properties of a polymer are greatly influenced by its topology as it has been proved already that glass transition temperature and melting temperature of a polymer is greatly altered by the introduction of branches in the structure.

For a new set of polymers, the main point concerning their thermal properties is its glass transition temperature. Glass transition in polymers is not a first order transition and is observed in the amorphous region of a polymer only. In this region, the long polymer chains are oriented randomly and have more freedom to move. The glass transition temperature is highly related to segmental chain motion, as polymers warm up. Glass transition temperature or in short T_g is greatly dependent on the structure of the polymers, the number of end groups, the polarity of the end groups and also on the number of cross-linked or branch points. Tg depends on the mobility and flexibility of the polymer backbone [34]. In addition, T_g corresponds to the free volume of the polymer which in turn is dependent on the end group interactions. T_g arises from the amorphous part of the polymer as they are semi crystalline in nature. Glass transition temperature decreases as the number of functional end groups increases while it increases by the introduction of more branch and crosslinking points [35].

As T_g is largely related to the large segmental motions in the polymer chain segments and it has been observed that above a certain molecular T_g is no further dependent on the terminal groups. In case of hyperbranched polymers, the determination of Tg becomes very much difficult because the segmental motion is greatly affected by the branching points and the presence of numerous end groups. Therefore in case of hyperbranched polymers, glass transition must be considered as a translational movement of the entire molecule instead of a segmental movement [36]. The chemical nature of the polymer also affects the T_g; for example, an aliphatic polyester generally has a much lower Tg than an aromatic one as examined by Voit and his group. They characterized a series of hyperbranched aromatic polyesters with different end groups. On conversion of carboxylic acid group to acetate end group, T_g shifted as much as 100 °C (from 250 to 150 °C) [37].

T_g is generally examined for most of the hyperbranched polymers, either calorimetrically or based on rheological measurements. Kim and his group characterized glass transition temperature of hyperbranched polyphenylenes and compared them with the small molecule counterparts. He concluded that HBPs had a similar relaxation mechanism with the small molecules [21]. They also studied the effect of terminal functional groups on the thermal relaxation behavior. Afterwards, Hawker and his group studied about the influence of functional end groups and DB on the thermal properties of the hyperbranched poly (ether-ketones) [35]. The glass transition results of these HBs showed that thermal stability of polymers was independent of macromolecular architecture, but depended largely on the nature of functional end groups. Again while characterizing hyperbranched polyglycerols and their esterified derivatives, Frey and his group further concluded hydrogen bonding between the terminal groups has a large effect on the glass transition temperatures of highly polar HBs [38].

In contrast, Jayakannan and his group tried to estimate the effect of branching on the T_g of a hyperbranched polymer. They synthesized different sets of poly (4-ethyleneoxyl benzoate)s having hyperbranched architecture differing in their DB. They observed that with increment in the branching units in the structure, there is a huge shift in the glass transition temperature. They concluded that the changes of free volume with the difference in DB and also the functional terminal groups may be simultaneously affecting the T_g which marks the difference between hyperbranched polymers and their linear analogs [39].

Yan and coworkers found an unusual dependence of T_g on DB of the hyperbranched polymer. They examined a series of hyperbranched poly[3-ethyl-3-(hydroxymethyl)oxetane] (PEHO) with varying DB. Other parameters such as molecular weight, molecular weight distribution, and the end group content were kept almost similar for all the sets. As shown in Fig. 4.6, T_g initially increases with increasing DB, passes through a maxima and finally decreases after DB crosses 27% [40].

The results can be analyzed by considering two factors: branching point increases with increasing DB making the structure more compact and rigid. Therefore, consequently T_g also increases. Again, with increasing DB, the free volume also increases which restricts the molecular mobility. Therefore after DB

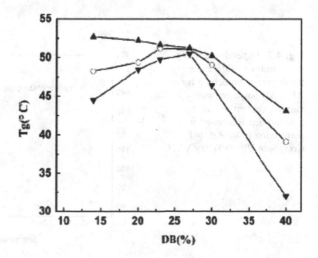

Fig. 4.6 Relationship between glass transition temperature (T_g) and degree of branching (DB) for poly [3-ethyl-3-(hydroxymethyl) oxetane]s. Reprinted with permission from Ref. [40]. Copyright (2009) American Chemical Society

crosses 27%, this factor predominates over the initial factor and thus there is a decrease in the glass transition temperature. Due to the synergistic effect of both the factors, the Tg passes through a maxima and then decreases sharply.

The end group also affects the glass transition strongly. With increasing number of polar end groups, interactions between the molecules increase and as a consequence Tg gets altered. As evident in Fig. 4.7, hyperbranched polyesters have been synthesized from AB_2 and AB type monomer with strong hydrogen bonding which increases with increasing DB and end groups. The hydrogen bonding affects the flexibility of the polymeric chains and thus the Tg changes as evident from the figure [41].

However, due to the combination of so many factors such as DB, end group polarity, backbone rigidity and steric interactions, which affects the glass transition temperature, it is still quite complicated to predict the Tg of a hyperbranched polymer using any standard model.

The crystallization and melting behaviors of polymers are also strongly influenced by the polymeric architecture. As for a basic example, with lesser branched structure HDPE shows high crystallinity with high melting temperature, whereas LDPE, containing random branches in its structure, shows lower melting temperature with lowering in crystallinity. Thus, it can be concluded that introduction of branches in the polymeric structure restricts proper alignment of the molecular chains and therefore hyperbranched polymers exhibit less crystallinity and lower melting point [42].

Yan et al. examined the different sets of hyperbranched polyethers synthesized from poly[3-ethyl-3-(hydroxymethyl)oxetane] (PEHO) having different DB [40]. While deviating from linearity, these polymers showed predominating amorphous characteristics. With increasing DB, there was a lowering in degree of crystallinity and melting temperature as well. Thus, branching lowers the crystalline domain in the polymeric structure. The dependence of degree of crystallinity with DB is given in Fig. 4.8.

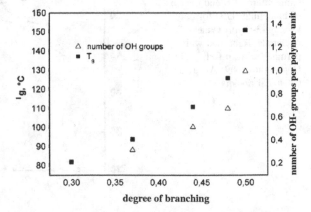

Fig. 4.7 Dependence of Tg and number of OH groups on the degree of branching for an AB_2+AB polyester with varying content of *AB* comonomer. Reprinted with permission from Ref. [41]. Copyright (2008) Wiley VCH

Fig. 4.8 Effect of DB on the crystallization behavior of PEHO. Reprinted with permission from Ref. [40]. Copyright (2009) American Chemical Society

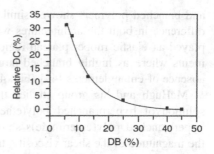

Thermal properties of polymers are characterized by instruments like Differential Scanning Calorimetry (DSC), Thermogravimetric Analysis (TGA), etc. The trivial nature of hyperbranched polymers is still under investigations and more detailed characterization.

4.6.2 Rheological Properties

Hyperbranched polymers show distinctive rheological properties which make it quite different from its linear analog. The rheological properties of a hyperbranched polymer make it more acceptable and open a wide range of application field for this special kind of architectural polymer. Therefore, these polymers are used as rheology modifiers, which when used in combination with a linear polymer, changes the processing properties of the same. Several research works have been done on the rheology of polymers, both theoretically and experimentally. The results revealed that the rheology of hyperbranched polymers deviates a lot from that of linear or star polymers. The main reason behind this behavior is the DB and the topology of the polymer. High DB and densely branched architecture restrict the hyperbranched polymers from entanglement. Due to lack of entanglements, hyperbranched polymers behave as Newtonian fluids [43]. Therefore, detailed knowledge about the rheology of hyperbranched polymers is required to understand the influence of DB, branching topology and also the effect of the end groups on the rheological properties of these type of architectural polymers.

For proper evaluation of the rheological properties, hyperbranched polymers have been synthesized having variation in DB, but the chemical composition was tried to be kept constant, so that the effect of the end groups on the rheological properties can be neglected. Ye and his group tried to study about the rheological behavior of polyethylene having different architecture varying from highly branched to nearly linear. Dynamic oscillations and steady shear measurements were carried out. Highly branched structures showed Newtonian flow behavior where as linear polyethylene showed typical shear thinning alike conventional polymers. The stress relaxation expressed by the storage (G') and the loss moduli (G'') were also examined. The results revealed that though at low-frequency region, both the linear and branched polymers showed similar characteristics but at higher frequency the difference in both the architectures was evident. The less branched polymers displayed an elastic rubber plateau region in the curve owing to its chain entanglements where as highly branched structures showed no such region proving the absence of entanglements [44] (Fig. 4.9).

McHugh and his group studied the rheological behavior of the concentrated solution of hyperbranched poly(etherimide) varying in their DBs [45]. These hyperbranched poly(etherimide)s showed characteristic Newtonian behavior, and the magnitude of the shear viscosity, the onset of the shear thinning, and the rise of normal stress effects directly correlated with the DB. Even the segment of mass between the branching points also affects the rheological properties of the hyperbranched polymers.

Pakula and his group examined hyperbranched poly(methyl methacrylates). Through rheological studies, it was evident that hyperbranched polymers have a broad relaxation behavior and the complex viscosity of the polymer behaves close to that for microgels and near-critical gels [46].

Kannan and his coworkers did extensive studies and compared the rheological properties of different architectural forms of polystyrene- linear, star and hyperbranched grade [47]. Different properties such as intrinsic viscosity ($[\eta]$), Rg, viscometric radius ($R\eta$), and zero-shear rate viscosity (η_0) were measured and compared. As already discussed earlier in this chapter, hyperbranched polystyrene exhibited lower intrinsic viscosity, Mark–Houwink exponent, and hydrodynamic radius when compared to the linear polystyrene of the same molecular weight. The experimental data suggested that hyperbranched polystyrene experienced steric hindrance due to its short branches and high DB and thus there was no chance of chain entanglements in the hyperbranched polymer, resulting in a linear dependence of η_0 versus molecular weight.

Johansson and his coworker have tried to establish relation between the branching extent in a hyperbranched polymer and its rheological property. They synthesized a series of hyperbranched aliphatic polyethers originating from 3-ethyl-3-(hydromethyl)oxetane with varying DB [48]. As observed from Fig. 4.10,

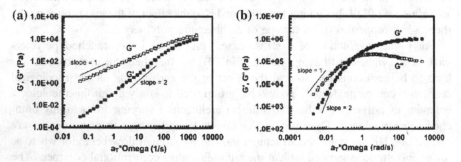

Fig. 4.9 Storage modulus G' and loss modulus G'' versus angular frequency at 60 °C for polyethylene with **a** highly branched topology and **b** nearly linear chain topology. Reprinted with permission from Ref. [44]. Copyright (2004) Wiley VCH

Fig. 4.10 Complex dynamic viscosity versus temperature for polyethers with different DBs. Reprinted with permission from Ref. [48]. Copyright (2002) Elsevier

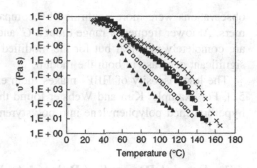

Fig. 4.11 Complex viscosity versus frequency for OH- and C12- terminated aromatic hb polyester. Reprinted with permission from Ref. [50]. Copyright (2000) Wiley VCH

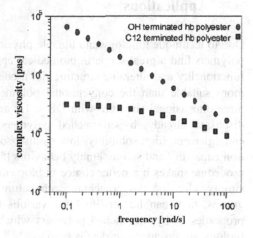

the set with highest DB (DB = 0.41) behaved as a completely amorphous material whose viscosity decreases rapidly above its Tg. Along with that, absence of any rubbery plateau region in the curve suggested that there is no chain entanglement present in the structure, whereas the set with lowest DB behaved as a semi crystalline polymer.

Adolf and group established the fact that for hyperbranched polymers a high shear rate is needed with increment in DB to initiate shear thinning for application purpose [49]. They used computer simulation to study the rheology of HBPs with different DBs under shear. They concluded that HBs with a low DB had a more scattered structure, while those with a high DB possessed a very compact structure. Therefore, the conformation change of HBPs under shear flow was quite different.

Apart from branching topology, the end groups also affect the rheological properties of a hyperbranched polymer. For example, Voit and his group studied the rheological properties of hyperbranched aromatic polyesters before and after modification with alkyl chains [50]. Figure 4.11 shows the complex viscosity dependence on the frequency for both polymers. The OH-terminated polymer showed nonentangled elastic behavior while the C^{12} alkyl chain modified form exhibited typical behavior for a viscous melt with a plateau at low frequencies. The observations were more confirmed by comparing the modulus for both the polymers. At lower frequency range though G' and G'' for the OH-terminated polyester are completely identical but for the modified hyperbranched polymer there was a significant decrease in both the moduli.

The low viscosity of HBPs makes them excellent modifiers and additives [51, 52]. For example, Kim and Webster found that the addition of a small amount of hyperbranched polyphenylene into polystyrene reduced the melt viscosity greatly.

4.7 Special Properties Related to Latest Biological Applications

Due to its unique topology and tunable physicochemical properties, hyperbranched polymers find a great scope in biomedical applications. Due to its high-end group functionality and alterable structure, its modification for biological applications is more suitable than the conventional polymers. Hyperbranched polymers have a three-dimensional structure with DB < 1.0 and high polydispersity index. And as discussed already, hyperbranched polymers also possess low molecular chain entanglement, high solubility, low melting/solution viscosity, excellent encapsulation capability and self-assembly behaviors [53]. Along with that ease of synthesis, procedure makes it a better choice in biomedical applications over dendrimers. By simply adjusting the branching architecture along with the terminal functional groups, they can be modified for various biological applications. The various properties of hyperbranched polymers which make them suitable in the field of biology are discussed in details below.

4.7.1 Biodegradability and Biocompatibility of Hyperbranched Polymers

The most important feature that makes hyperbranched polymers suitable for biomedical application is the synthesis of biodegradable and biocompatible backbone. Hyperbranched polymers with controlled DB, having proper degradation kinetics, and excellent biocompatibility are quite suitable as drug delivery systems, scaffolds for tissue engineering, imaging agents, and other biomaterials [54, 55]. Hyperbranched polymers from polyesters, polypeptides, polyethylene oxides, polyglycerols, and polyphosphates have been synthesized and widely used in the biological field. In order to meet the increasing demand for more variety of biocompatible hyperbranched polymers, researchers all around the world are trying to develop new materials with enhanced physicochemical properties.

Initially, polymers with high biocompatibility were developed. For example, Frey and his coworkers synthesized polyglycerols via anionic ring-opening multibranching polymerization from AB_2 type monomer. The controlled polymerisation leads to hyperbranched polyglycerols with excellent chemical stability, high hydrophilicity and excellent biocompatibility. Thus, the material became a obvious choice in the biological and medical field. But the main disadvantage with the material was that it was not biodegradable. The constant accumulation of polyether from the hyperbranched polyglycerol backbone was injurious for living cells [56]. Thus the requirement for biocompatible polymer became important that can degrade in physiological condition after fulfilling its purpose. Therefore, hyperbranched polyesters gained immense importance due to its chemical structure which can undergo hydrolytic degradation easily. The most common hyperbranched polyester used commercially is Boltron Hx (x = 20, 30, 40).

Moreover, stimuli responsive hyperbranched polymers have also gained attention in the field of biomedicals. Functional groups sensitive to thermal condition, pH, redox condition can be incorporated within the structure making the polymer more action specific [57].

Kizhakkedathu and coworkers modified the hyperbranched polyether synthesized via ring-opening multibranching polymerization and incorporated ketal groups into the backbone of hyperbranched polyether to induce biodegradability [58]. Both cyclic and acyclic ketal groups were incorporated in the structure to generate pH responsive PKHEs. Due to the presence of both type of ketal group in the structure, the degradation kinetics of modified PKHEs could be could be altered from a few minutes to a few hundred days by just adjusting the pH of the medium (Fig. 4.12). In addition to that, another advantage is that PKHEs and their degradation products exhibited high biocompatibility as demonstrated by the cell viability assay and blood compatibility assay results, and thus these hyperbranched PKHEs show great potential as multifunctional drug delivery vehicles.

Even hyperbranched peptides have gained more importance in the biomedicals as they generally displayed higher solubility, enhanced proteolytic stability and a

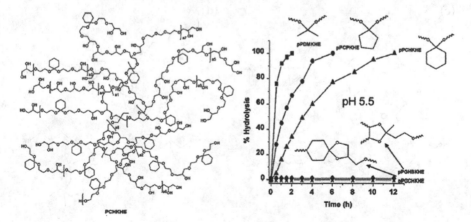

Fig. 4.12 Schematic structures of hyperbranched PKHEs and comparison of hydrolysis kinetics plots of different PKHEs at pH 5.5. Reprinted with permission from Ref. [58]. Copyright (2012) American Chemical Society

lower toxicity than their linear analogs. Even hyperbranched polypeptides have proven to be a great alternative for their symmetrical analog as peptide dendrimers are more costly with difficult synthesis procedure and therefore difficult to be commercialized [59]. Recently, these hyperbranched polypeptides have triggered the interest among the researchers for development of various synthetic strategies for hyperbranched polypeptides and their potential bioapplications can be investigated thoroughly.

The biocompatibility and biodegradability of hyperbranched polyphosphates were studied by Huang and Yan. They used hydroxyl-functionalized cyclic phosphate inimer 2-(2 hydroxyethoxy) ethoxy-2-oxo-1,3,2-dioxaphospholane (HEEP) and through self-condensing ring-opening polymerization they prepared an eminent biocompatible and biodegradable polymer which can find a significant application in the field of biomedicals [60] (Fig. 4.13).

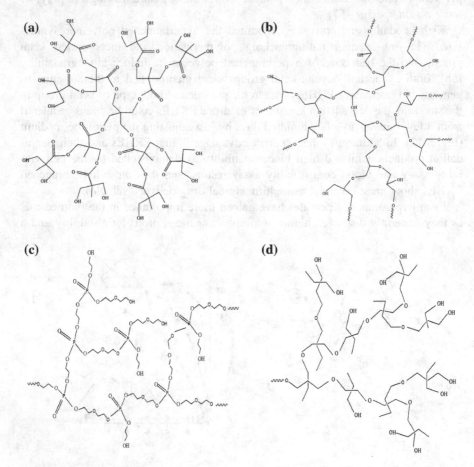

Fig. 4.13 Schematic structures of hyperbranched polymers used for biomedical applications: **a** polyesters, **b** polyglycerols, **c** polyphosphates, **d** hyperbranched poly (3-ethyl-3-oxetanemethanol)

4.7.2   Self-Assembly of the Hyperbranched Polymers

Dendrimers and conventional linear polymers, surfactants show self-assembly mechanism quite easily due to its structural ease. But the phenomenon of self-assembly in hyperbranched polymers with randomly branched structure was quite unique. But the research done by Yan and his coworkers in 2004, triggered the interest of all the polymer researchers in this sector of hyperbranched polymers. They reported that supramolecular self-assembly structure was obtained from amphiphilic HBPO-star-PEO with a hydrophobic hyperbranched poly (3-ethyl-3-oxetanemethanol) core (HBPO) and many hydrophilic poly(ethylene oxide) arms (PEO) in acetone [61]. This observation initiated the wide research in this field. Researchers and scientists all over the field tried to develop more hyperbranched polymers with self-assembling nature that can be widely used in the biomedicals as encapsulators or delivery vehicle.

Supramolecular self-assembly as obtained from hyperbranched polymers has unique and special characteristics which differ largely from those obtained from conventional polymers. Hyperbranched polymers have varying topology which can be controlled through its molecular weight and DB. Thus, by varying these two factors different supramolecules can be obtained with different dimensions. Then the assemblies can form unimolecular micelle with dimension range of about 10 nm, much smaller than that of the conventional polymers. Due to its globular structure, the mechanism of forming self-assembly is also different in case of hyperbranched polymers. Hyperbranched polymers can form self-assembly in different shapes and scales, such as spherical micelles, nano- or micro-scale vesicles, ribbons, honeycomb-patterned films, fibers, tubules, etc. (as given in Fig. 4.14) [62].

Thus it can be concluded that the self-assembly characteristics of hyperbranched polymers pave a new path to be investigated further and design various molecules for biomedical applications.

4.7.3   Encapsulation by Hyperbranched Polymers

Encapsulation phenomenon means incorporation of an active substance in the matrix of a carrier component. Hyperbranched polymers have large void spaces within its structure making it a potential material for host–guest applications. Therefore in the field of biomedicals, hyperbranched polymers are a great option as encapsulators for both hydrophillic and hydrophobic materials [63]. Encapsulation can be via two mechanism: may be the drugs can get into the cavity within the voids present within the hyperbranched polymer structure and get entrapped by physical interaction [64], or may be the drug gets within the unimolecular micelle formed by the self-assembly of the hyperbranched polymers and gets more stabilized than multimolecular micelle formed from other conventional amphiphilic polymers (Fig. 4.15).

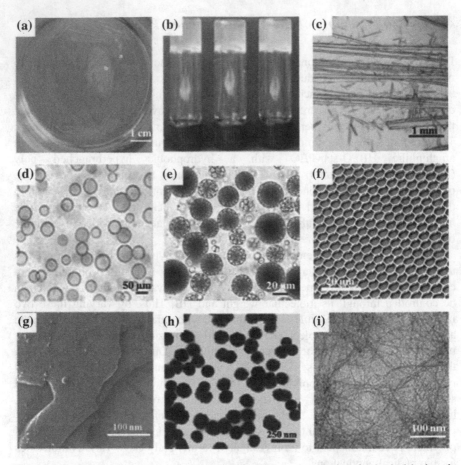

Fig. 4.14 Self-assembled structures of amphiphilic HBPs: **a** macrotubes, **b** physical hydrogel, **c** mesoscopic tubes, **d** giant vesicles, **e** composed vesicles, **f** honeycomb films, **g** 2D sheets, **h** spherical micelles, **i** nanoscale fibers. Reprinted with permission from Ref. [62]. Copyright (2015) Royal Society of Chemistry

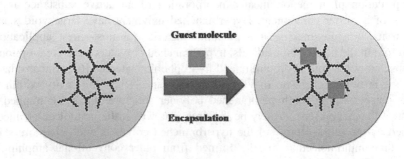

Fig. 4.15 Encapsulation mechanism by hyperbranched polymers

Encapsulation in hyperbranched polymers occurs via physical, chemical or high-pressure technique depending on the size of the active substance and also on the pore size of the polymers. The different techniques of encapsulation within the polymer matrix are demonstrated in the scheme given below (Fig. 4.16).

The physical method of encapsulation means mechanical techniques in which an active substance is either coated or physically trapped within the carrier film, where as in chemical method, the hyperbranched polymer is dissolved in a solvent which forms a shell or matrix around the active substance chemically and entraps it [65].

In molecular encapsulation process, hyperbranched polymers form a self-assembly structure to entrap guest molecules. The most important factors that lead to self-assembly are hydrogen bonds and metal–ligand interactions. The subsequent encapsulation of the guest molecule is dependent on the size and shape of the pores provided by the guest molecule, as well as on its chemical interaction with the macromolecule and the solvent [66].

Hyperbranched polymers have high PDI with a wide range of molar masses. The polydispersity in terms of both molar and particle size greatly affects the encapsulation process. The variation of molar mass affects the solubility/viscosity of the polymers and thus it affects the end performance of the polymer as vehicles for delivery purpose. The variation of pore size affects the release kinetics in hyperbranched polymers which may produce undesirable variation in drug stability and drug release profiles which limits the application of hyperbranched polymers for pharmaceuticals. Thus, researchers are trying to develop hyperbranched polymers with monodisperse characteristics, so that the encapsulation and release

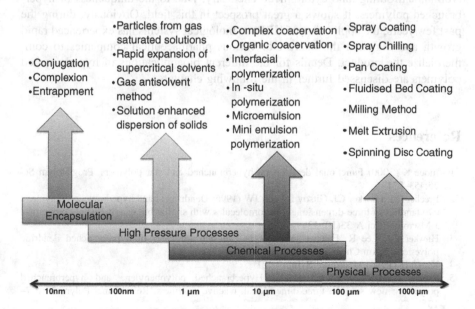

Fig. 4.16 Encapsulation mechanism at different microparticle size

mechanisms are more controlled. Further investigations must be carried out to commercialize the hyperbranched polymers more in this field.

4.8 Conclusion

In this chapter we have discussed the structure–property relationship of hyper-branched polymers and also the properties that make hyperbranched polymers suitable for bioapplications are also discussed. DB is the most important parameter for hyperbranched polymers. The solution properties, thermal properties, rheological properties depend largely on this branching percentage. As elaborated in this chapter, it is evident that different structural hyperbranched polymers with different tunable properties. In the last two decades, the researches ongoing on hyper-branched polymers have opened a vast field for architectural polymers. By controlling the DB, the properties of hyperbranched polymers can be altered making it a versatile material. The terminal functional groups also play a major role in property alteration for hyperbranched polymers. Presence of polar end groups affects the solution properties, thermal properties, rheological properties as well as biological properties.

The potential for HBPs in biological and biomedical applications is immense as discussed in this chapter. Therefore, hyperbranched polymers find applications in various biomedical purposes such as in therapy (drug delivery, gene transfection, and protein delivery), bioimaging, biomineralization, tissue engineering, antimicrobial, antifouling, and cytomimetic chemistry. Due to the uniqueness of hyper-branched polymers, it shows a great prospect in this field. Obviously, during the past few years, the application of HBPs in biological systems has experienced rapid growth and triggered the interest of various pharmaceutical companies to commercialize the product. Details for the different bioapplications of hyperbranched polymers are discussed further in the following chapters.

References

1. Inoue K (2000) Functional dendrimers, hyperbranched and star polymers. Prog Polym Sci 25:453–571
2. Fréchet JM, Hawker CJ, Gitsov I, Leon JW (1996) Dendrimers and hyperbranched polymers: two families of three-dimensional macromolecules with similar but clearly distinct properties. J Macromol Sci A 33(10):1399–1425
3. Hawker CJ, Lee R, Fréchet JMJ (1991) One-step synthesis of hyperbranched dendritic polyesters. J Am Chem Soc 113:4583–4588
4. Kim YH (1994) Macromol Symp 77:21
5. Fréchet JMJ, Hawker CJ (1995) Hyperbranched polyphenylene and hyperbranched polyesters: new soluble, three-dimensional, reactive polymers. React Funct Polym 26:127–136

6. Flory PJ (1953) Principles of polymer chemistry. Molecular weight distribution in non-linear polymers and the theory of gelation, Chapter 9, 348. Cornell University Press, Ithaca

7. Kim YH, Webster OW (1990) Water soluble hyperbranched polyphenylene: "a unimolecular micelle?" J Am Chem Soc 112:4592

8. Hölter D, Burgath A, Frey H (1997) Degree of branching in hyperbranched polymers. Acta Polym 48(1–2):30–35

9. Hölter D, Fret H (1997) Degree of branching in hyperbranched polymers. 2. Enhancement of the DB: scope and limitations. Acta Polym 48.8:298–309

10. Frey H, Hölter D (1999) Degree of branching in hyperbranched polymers. 3 Copolymerization of ABm-monomers with AB and ABn-monomers. Acta Polym 50 (2-3):67–76

11. Galina H, Walczak M (2005) A theoretical model of hyperbranched polymerization involving an ABf monomer. Polimery 50(10)

12. Maier G, Zech C, Voit B, Komber H (1998) An approach to hyperbranched polymers with a degree of branching of 100%. Macromol Chem Phys 199:2655–2664

13. Hobson LJ, Kenwright AM, Feast W (1997) A simple 'one pot' route to the hyperbranched analogues of Tomalia's poly(amidoamine) dendrimers. J Chem Commun 19:1877–1878

14. Merino S, Brauge L, Caminade AM, Majoral JP, Taton D, Gnanou Y (2001) Synthesis and reactivity of small phosphorus-containing dendritic wedges (dendrons). Chem—Eur J 7:3095

15. Huber T, Böhme F, Komber H, Kronek J, Luston J, Voigt D, Voit B (1999) Macromol Chem Phys 200:126

16. Komber H, Stumpe K, Voit B (1814) Macromol Chem Phys 2006:207

17. Kambouris P, Hawker CJ (1993) J Chem Soc Perkin Trans 1, 22:2717

18. Bolton DH, Wooley KL (2002) J Polym Sci Part A: Polym Chem Ed, 40:823

19. Sheridan PF, Adolf DB, Lyulin AV, Neelov I, Davies GR (2002) Computer simulations of hyperbranched polymers: the influence of the Wiener index on the intrinsic viscosity and radius of gyration. J Chem Phys 117(16):7802–7812

20. Hult A, Johansson M (1999) Malmstr"om. E Adv Polym Sci 143:1

21. Kim YH, Webster OW (1992) Hyperbranched polyphenylenes. Macromolecules 25 (21):5561–5572

22. Fox TGJ, Flory JP (1950) J Appl Phys 21:581

23. FrCchet JMJ (1994) Presented at the 35th IUPAC International symposium on macromolecules. Akron, Ohio

24. Kainthan RK, Brooks DE (2007) In vivo biological evaluation of high molecular weight hyperbranched polyglycerols. Biomaterials 28(32):4779–4787

25. Turner SR, Voit BI, Mourey TH (1993) Macromolecules 26:4617

26. Voit BI, Lederer A (2009) Chem Rev 109:5924

27. Lederer A, Abd Elrehim M, Schallausky F, Voigt D, Voit B (2006) e-Polym 039

28. Yan D, Zhou Z (1999) Molecular weight distribution of hyperbranched polymers generated from polycondensation of AB2 type monomers in the presence of multifunctional core moieties. Macromolecules 32(3):819–824

29. Yan D, Zhou Z, Müller AH (1999) Molecular weight distribution of hyperbranched polymers generated by self-condensing vinyl polymerization in presence of a multifunctional initiator. Macromolecules 32(2):245–250

30. Gittins PJ, Alston J, Ge Y, Twyman LJ (2004) Macromolecules 37:7428

31. Radke W, Litvinenko G, Mu"ller AHE (1998) Macromolecules 31:239

32. Burchard W (1999) Adv Polym Sci 143:113

33. Lederer A, Boye S (2008) LCGC Ads 24

34. Markoski LJ, Moore JS, Sendijarevic I, McHugh AJ (2001) Macromolecules 34:2695

35. Hawker CJ, Chu FK (1996) Macromolecules 29:4370

36. Sperling LH (1986) Introduction to physical polymer science. Wiley, New York Chapter 6

37. Voit BI (1995) Acta Polym 46:87

38. Sunder A, Bauer T, M"ulhaupt R, Frey H (2000) Macromolecules 33:1330

39. Jayakannan M, Ramakrishnan S (2000) J Polym Sci, Part A: Polym Chem 38:261

40. Zhu Q, Wu JL, Tu CL, Shi YF, He L, Wang RB, Zhu XY, Yan DY (2009) J Phys Chem B 113:5777
41. Schallausky F, Erber M, Komber H, Lederer A (2008) Macromol Chem Phys 209:2331
42. DeSimone JM (1060) Science 1995:269
43. Farrington PJ, Hawker CJ, Fréchet JMJ, Mackay ME (1998) Macromolecules 31:5043
44. Ye Z, Alobaidi F, Zhu S (2004) Macromol Chem Phys 205:897
45. Sendijarevic I, McHugh AJ (2000) Macromolecules 33:590
46. Simon PF, Müller AH, Pakula T (2001) Macromolecules 34:1677–1684
47. Kharchenko SB, Kannan RM (2003) Macromolecules 36:399
48. Magnusson H, Malmström E, Hult A, Johansson M (2002) Polymer 43:301
49. Lyulin AV, Adolf DB, Davies GR (2001) Macromolecules 34:3783
50. Schmaljohann D, Häußler L, Pötschke P, Voit BI, Loontjens TJA (2000) Macromol Chem Phys 201, 49
51. Jannerfeldt G, Boogh L, Manson J-AE (1999) J Polym Sci, Part A: Polym Chem 37:2069
52. Star A, Stoddart JF (2002) Macromolecules 35:7516
53. Sharma K, Zolotarskaya OY, Wynne KJ, Yang H (2012) J Bioact Compat Polym 27:525
54. Xu S, Luo Y, Haag R (2007) Macromol Biosci 7:968
55. Zhou Y, Huang W, Liu J, Zhu X, Yan D (2010) Adv Mater 22:4567
56. Calderon M, Quadir MA, Sharma SK, Haag R (2010) Adv Mater 22:190–218
57. Hu M, Chen MS, Li GL, Pang Y, Wang DL, Wu JL, Qiu F, Zhu XY, Sun J (2012) Biomacromolecules 13:3552–3561
58. Shenoi RA, Narayanannair JK, Hamilton JL, Lai BF, Horte S, Kainthan RK, Varghese JP, Rajeev KG, Manoharan M, Kizhakkedathu JN (2012) J Am Chem Soc 134:14945–14957
59. Chang X, Dong CM (2013) Biomacromolecules 14:3329–3337
60. Liu JY, Huang W, Zhou YF, Yan DY (2009) Macromolecules 42:4394–4399
61. Yan DY, Zhou YF, Hou J (2004) Science 303:65–67
62. Wang D, Zhao T, Zhu X, Yan D, Wang W (2015) Chem Soc Rev 44:4023–4071
63. Shi YF, Tu CL, Wang RB, Wu JL, Zhu XY, Yan DY (2008) Langmuir 24:11955
64. Zou J, Shi W, Wang J, Bo J (2005) Macromol Biosci 5:662–668
65. Ye L, Letchford K, Heller M, Liggins R, Guan D, Kizhakkedathu JN, Brooks DE, Jackson JK, Burt HM (2010) Biomacromolecules 12:145–155
66. Kainthan RK, Mugabe C, Burt HM, Brooks DE (2008) Biomacromolecules 9:886–895

Chapter 5
Latest Biomedical Applications of Hyperbranched Polymers: Part 1: As Delivery Vehicle

5.1 Introduction to the Concept of Targeted Delivery

Alike dendrimers, hyperbranched polymers also play a significant role in various biomedical applications. An ideal delivery vehicle in biological applications should possess the following characteristics: excellent biocompatibility and biodegradability, it must have the ability to form a stable complex with the external agent, transport the external agent into the specified targeting site, and then release them in a controlled manner, while the physiological properties of the agent is kept intact. Polymer-based biomaterials are thus quite preferred as delivery vehicles over other small molecules as it meets all the above mentioned criteria. Again among all the polymers, hyperbranched polymers are considered as an ideal matrix for drug and gene delivery vehicles because of its tailorable architecture and availability of plenty of functional groups. Along with that, the typical characteristics of hyperbranched polymers like low (melt) viscosity and compact structure also contributes towards the more demand for hyperbranched polymers in the biological field. An additional promising feature is their void structure which has the ability to act as a host for smaller molecules in various ways. The main biomedical fields in which hyperbranched polymers are likely to find applications are gene delivery, drug delivery, protein delivery, as biodegradable materials, modification of surfaces (biocompatibilization, antifouling, etc.) and bulk materials, as bioimaging agent and applications related to biointeractions. With the passing years, researchers are trying to explore the use of hyperbranched polymers as drug carriers for the efficient and controlled delivery of drugs to the specified targeted sites. Although similar applications have been suggested and explored for dendrimers, their high cost and tedious synthetic procedure will favors the application of hyperbranched polymers over their symmetrical analogs. Many of the suggested potential applications, however, are still left to be explored till date [1, 2]. The encapsulation studies for both hydrophilic and hydrophobic guest molecules using hyperbranched polymers such as hyperbranched polyesters, polyglycerols, polyesteramides, and

A. Bandyopadhyay et al., *Hyperbranched Polymers for Biomedical Applications*, Springer Series on Polymer and Composite Materials, https://doi.org/10.1007/978-981-10-6514-9_5

polyethylenimines has been extensively explored by the researchers in the past few years as already discussed in the previous chapters [3–5].

5.2 Encapsulation Ability of Hyperbranched Polymers

The primary feature that enables hyperbranched polymer to be used as delivery vehicle is its excellent encapsulating capability. The hydrophilic and hydrophobic guest molecules get entry within the hyperbranched polymers via two processes. First, the drug gets encapsulated within the cavities present within the hyperbranched structure via simple physical interactions [5]. Second, drugs can get entrapped within the unimolecular micelle formed by the self assembling of the hyperbranched polymers. It has been already stated that the unimolecular micelle formed by the hyperbranched polymers display better stability than multimolecular micelles formed from the conventional amphiphilic copolymers under similar standard conditions [6].

Haag and coworkers tried to get a comparison of encapsulating ability between hyperbranched and linear polymers. Hyperbranched PEI was modified with different functional fatty acids (C18, C16, C11, C6) and the resulting modified polymers were ampiphillic in nature. The structures were based on core–shell architectures which was capable of entrapping anionic guest molecules such as suitable caboxylate, sulfonate, phosphate, and acidic OH groups (Fig. 5.1). They observed that the fatty acid modified PEI exhibited much higher encapsulating capacity than their linear analogs [7].

Yan and coworkers have prepared a phospholipid mimicking material based on HPEEP [8]. Self condensing ring opening polymerization of hydroxyl functionalized cyclic phosphates and then the polymer was further modified with palmitoyl chloride which leads to a ampiphillic polymer with a hydrophilic HPHEEP head and plenty hydrophilic alkyl tails. These amphiphilic hyperbranched polymers had the capability of self-assembling in aqueous media. The micelles were in nano-range of size and thus can easily get entry into the living cells. An in vitro investigation was done using a hydrophobic anticancer drug, chlorambucil, loaded within the hyperbranched polymer, MCF-7 breast cancer cells. The results showed that there was a restriction in the rapid growth of the cancer cells (Fig. 5.2).

However, the main drawback with the unimolecular micelles of hyperbranched polymers is that due to its restricted extent of interior cavities, it is allows only a small amount of guest molecules to be entrapped within the matrix. Therefore multimolecular micelles from hyperbranched polymers with larger hydrophobic cores, higher encapsulating capacity and better controlled mechanism for drug delivery, are gaining more attention recently. Cheng and coworkers studied the drug loading capacity of a multimolecular micelle formed from the hyperbranched copolymer, H40-*star*-(PCL-*b*-PEG), attached with folate moieties used as the targeting ligands. Two antineoplastic drugs, 5-fluorouracil and paclitaxel were used in the study. In multimolecular micelles, the drugs got enfolded within the matrix

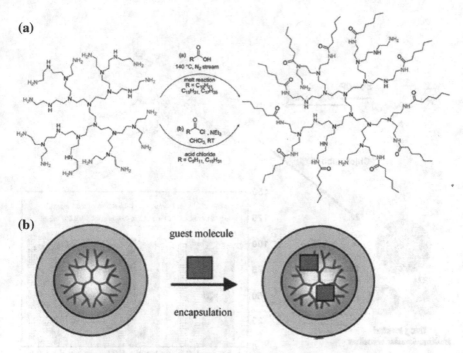

Fig. 5.1 a Modification of hyperbranched PEIs with different fatty acids leading to a ampiphillic copolymer. **b** Schematic representation of the core shell structure formed from the hydrophilic PEI and hydrophobic fatty acids which is capable of encapsulating polar guest molecules. Reprinted with permission from Ref. [7]. Copyright (2007) Wiley

through the non-covalent interactions. The results of the studies showed that the drug loaded micelle were very good biocompatible and is capable of restricting the abnormal growth of tumor cells [9].

Gao and coworkers studied about the multiple guest molecule encapsulation by hyperbranched polymers [10]. They observed that the loading capacity of one kind of guest molecule is enhanced in presence of other guest molecules within the amphiphilic hyperbranched polymers in the process of double-dye host–guest encapsulation. Thus it was concluded that hyperbranched polymers are also capable of multiple guest encapsulation instead of just one.

All the above-discussed topics demonstrate that the three-dimensional architecture of hyperbranched polymers plays a critical role in their encapsulation capacity but there are still some limitations which restrict the methodical and systematic drug encapsulation and controlled release phenomenon. The main disadvantage is due to non-covalent interaction, there is rapid and uncontrolled drug release from the hyperbranched core where as the covalently bound HB–drug conjugates exhibits better stability in both water and buffered solutions and the linker group used for the conjugates significantly affects the drug release

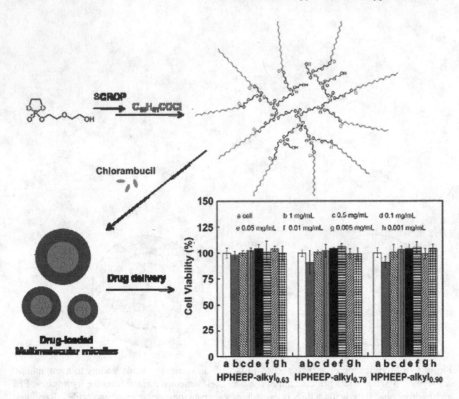

Fig. 5.2 A graphical representation of synthesis of phospholipid mimicking HPHEEP-alkyls and its self assembling capability. Reprinted with permission from Ref. [8]. Copyright (2010) Elsevier

kinetic [11]. The linker groups generally used for HB-drug conjugates must be easily biodegradable such as ester groups, acylhydrazone groups, or disulfide groups [12, 13].

5.3   Hyperbranched Polymers in Controlled Drug Delivery

The efficiency of the delivery vehicle is very much dependent on the release kinetics as demonstrated by the delivery system in the body. The release kinetics of drugs is of primary significance in a functional drug delivery system because it allows the vehicle to retain the drug content until the target site is reached and then goes for sustain release of the drugs for more effective action. Hyperbranched polymers exhibits high stability as drug carriers as it forms unimolecular micelles or self-assemble into multimolecular micelles. Moreover, hyperbranched polymers have many terminal functional groups, an alterable substitutional groups, and adjustable degree of branching to allow more regulated drug release.

Burt and coworkers worked on two commercially available hyperbranched polymers (HPG and H40) and modified them with carboxylic acid. These modified hyperbranched polymers were used as vehicle for the controlled release of anti-cancer drug cisplatin [14]. The drug delivery efficiency of both the polymers was examined thoroughly. It was observed that in physiological condition, the modified HPG formed strongly bound complexes with the drug; therefore there was a sus-tained release of the entire cisplastin drug over 7 days. But in case of H40 it was found that there was a rapid release of the drug which implied that the matrix formed higher proportion of weakly bound complexes with cisplatin. Though, after 5 days it was found that only 60% of the drug has released from the polymer matrix implying that the rest of 40% of the drug is strongly bounded with H40. Thus from the results it can be concluded that the complexes of HPG and cisplatin can be considered as promising delivery systems.

Further scientists have been trying to develop stimuli responsive hyperbranched for better controlled release applications. The polymer matrix is designed in such a manner that a slight change in pH, temperature, light, or redox conditions, greatly affects the sustained release of the drug at the target site.

Gong and coworkers constructed a biocompatible and biodegradable hyper-branched copolymer, H40-star-(PLA-b-PEG), as a vehicle for tumor-targeting drug (Fig. 5.3a) [15]. Folic acid was attached as targeting moieties. The anticancer drug DOX (Doxorubicin) got attached with the hydrophobic PLA blocks by pH-sensitive hydrazone linkages and thus in acidic medium, due to cleavage of hydrazone linkages, the rate of drug release significantly increased (Fig. 5.3b).

Ji and coworkers have tried to develop photoresponsive hyperbranched polymer. They modified the biocompatible and biodegradable HPHEEP with a photore-sponsive segment (Fig. 5.4), hydrophobic 2-diazo-1,2-naphthoquinone-5-sulfonyl chloride (DNQ) [16]. The resultant terminal-modified HPHEEP is capable of forming micelle in water and have excellent biocompatibility. Under UV irradia-tion, the DNQ moieties disintegrates and the micelle gets destabilized for triggered drug release.

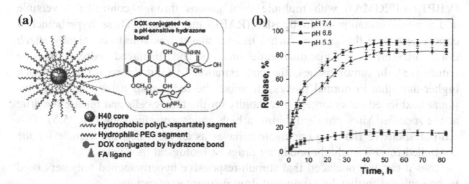

Fig. 5.3 **a** Schematic representation of the H40-star-(PLA-DOX)-b-PEG-OH/FA copolymer. **b** Release profiles of DOX from the H40-star-(PLA-DOX)-b-PEG-OH/FA micelles at 37 °C. Reprinted with permission from Ref. [15]. Copyright (2009) Elsevier

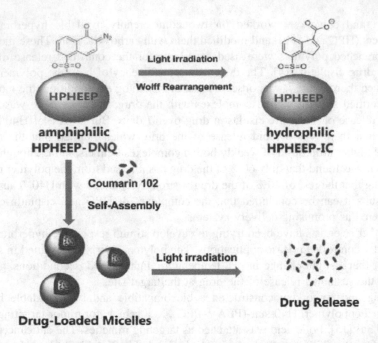

Fig. 5.4 Illustration of the self-assembly and photoresponsive behavior of HPHEEP-DNQ Reprinted with permission from Ref. [16]. Copyright (2011) Royal Society of Chemistry

Wang and his coworkers synthesized thermoresponsive PEG based hyperbranched copolymers using high concentration of ethylene glycol dimethacrylate as the branching agent via atom transfer radical polymerization (ATRP). These copolymers exhibited lower critical solution temperature (LCST) close to human body temperature [17].

Zhu and his coworkers have synthesized hyperbranched poly-((S-(4-vinyl) benzyl S0-propyltrithiocarbonate)-co-(poly(ethyleneglycol)methacrylate)) (poly (VBPT-co-PEGMA)) with multiple thiol groups through controlled reversible addition-fragmentation chain transfer (RAFT) mechanism. These hyperbranched copolymers were designed in such a manner that it was able to covalently attach thiol-containing drugs via disulphide linkages and displayed redox-responsive nature [13]. In tumor tissues, the concentration of Glutathione (GSH) is much higher than that in normal tissues, therefore the redox-responsive HB–drug conjugates exhibited better targeting capability for the tumor cells and release the drug at the specified sites, thus exhibiting a high anticancer efficiency (Fig. 5.5). This redox-responsive HB-drug conjugate behaves as an ideal delivery vehicle for the controlled release of thiol-containing drugs or biological molecules.

Thus it can be concluded that stimuli-responsive hyperbranched polymers need to be explored further for controlled drug delivery applications.

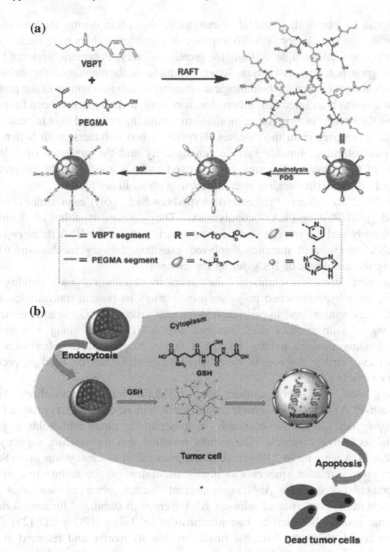

Fig. 5.5 Schematic illustration of **a** Synthesis of poly(VBPT-co-PEGMA)-S-S-MP; **b** redox-responsive behavior of poly(VBPT-co-PEGMA)-S-S-MP micelles for sustained drug release. Reprinted with permission from Ref. [13]. Copyright (2014) American Chemical Society

5.4 Hyperbranched Polymers in Protein Delivery

In the recent years, biologists have been trying to treat deep rooted clinincal disorders such as cancer, anemia and other metabolic diseases, by transferring active peptides and proteins in the targeted site [18]. But the hydrolytic instability and low bioavailability of the proteins is restricting the application. They can easily disintegrate by physical, chemical or enzymatic disruption during the process of peptide storage or delivery [19]. To improve on these researchers have been trying to develop new and unique systems for protein delivery. Polymeric systems have showed great potential as protein delivery vehicle as they enhance the stability, improves absorption across the biological boundaries, and also enhances the protein residence in the bloodstream. Furthermore, some polymers have also been found to improve the protein incorporation mechanism within the cells and thus increases its potential to recover from the diseases. Hyperbranched polymers, with better stability and solubility, tunable surface functionality and the presence of a large number of functional end-groups, have proved to be a good vehicle for protein transfer in the specified region in comparison with its linear grade.

Wu and coworkers synthesized hyperbranched poly[(ester-amine)-co-(D, L-lactide)] (HPEA-co-PLA) copolymers. They were capable of forming self-assembly and the micelle was used as protein carriers for BSA delivery [20]. The BSA-encapsulated micelles displayed excellent delivery mechanism while retaining the properties of BSA during the process.

Frey and coworkers utilized the excellent solubility and stability of biocompatible hyperbranched polyglycerols to study its protein transmission efficiency. They synthesized hyperbranched-b-linear (HPG-b-PEG) heterotelechelics consisting of a linear PEG block attached with a HPG block along with biotin. Biotin binding protein, avidin, formed a stable conjugated complex with the polymer and further the results have proved that HPG can be used as protein delivery vehicle [21].

Zhang and coworkers studied on polylactic acid functionalized HPG (HPG-star-PLA) as delivery vehicle for bovine serum albumin (BSA) protein [22]. The copolymer formed self-assembly in nanoparticle range and holds a great potential as delivery vehicle. The results revealed that the loading capacity of HPG-star-PLA were up to 23% where as the association efficiency was up to 86%, and the protein release kinetics was highly dependent on the architecture of the hyperbranched polymers. Another important feature observed was that the physicochemical properties of released BSA were well stabilized for over 4 days. The same group also tried to transport insulin by using HPG-g-CD [23]. The polymer could efficiently load the insulin within its matrix and released at the specific site which was evident from the decreased glucose level in the specimen.

As hyperbranched polymers exhibit an easy strategy for introduction of different terminal functional groups, it has the capability of covalently conjugating a variety of proteins and peptides. Klok and coworkers tried to study on this property further and thus they attached both BSA and lysozyme into the HPG and hyperbranched-linear polymer HPG-PEG with the squaric acid mediated coupling strategy, and subsequently used for protein delivery.

5.5 Hyperbranched Polymers in Gene Delivery

Gene therapy is used to treat previously incurable diseases with genetic disorders and is considered much better than the conventional treatment procedures. It involves transference of exogenous nucleic acids into the nucleus of the specific cells of the human body [24]. An effective gene delivery vehicle is required for proper transfer of genetic materials as they are easily degraded by serum nucleases in the blood when injected intravenously.

Nonviral polycationic vehicles are regarded as ideal gene delivery systems, because of its improved stability, enhanced biocompatibility and biodegradability, high flexibility of trans-gene size, low cost of synthesis, and easy scale-up procedure. The electrostatic interaction between the cationic polymeric vectors and the negatively charged DNA improves cellular uptake efficiency and transfection efficiency of DNA or RNA. With the advantage of its architecture, cationic hyperbranched polymers have gained attention as gene transfer vehicle. Polyamines like hyperbranched polyethylenimine (HPEI) and hyperbranched polypropylenimine (HPPI) have proved to be an efficient gene transfection agent. The availability of plenty of terminal amino groups with flexible molecular architecture increases the efficiency of these hyperbranched polyamines for more effective and specific gene transfer. HPEI displays several attractive features that makes it superior transfection agent in various cell lines and tissues. The presence of numerous terminal primary amines helps them to form nanosized compact structure by condensing nucleic acids within its matrix and transferring it to the specified cell and also resists the degradation of the DNA complexes during the transfer process. But the main drawback with HPEI is its high cytotoxicity and lack of cell specificity which limits further applications. Therefore researchers tried to modify HPEI to reduce cytotoxicity in the vehicle as well as increase the gene delivery efficiency. The modification of the surface of HPEI with functional small molecules, hydrophilic or biodegradable polymers, such as PEG, natural glucose polymers, proteins, peptides, etc., is done to improve the efficiency of the polymer. It was observed that the modified hyperbranched polyamines showed improved transfection efficiency than the unmodified polyamines. The transfer phenomenon and the cell specificity of the vehicle greatly depend on the functional group attached and also on the degree of substitution. For instance, peptide-modified HPEIs show better prosperity as gene transfer vehicle as they improve the biocompatibility and are easily metabolized [25].

Wu and coworkers modified HPEI (25 kDa) with histidine based peptide for gene transfection studies [26]. The polymer matrix showed improved transfection efficiency and enhanced cell survival to the human embryonic kidney cell line (HEK 293FT). Along with that, these modified HPEI could also act as a gene delivery vehicle for highly resistant cells such as human adipose stromal cells (ASCs), dermal fibroblasts, and cardiac progenitor cells (CPCs).

To reduce the cytotoxicity and also to improve the controlled release of the nucleic acids, the backbone of hyperbranched polyamines are generally modified by linking with biodegradable linkages such as disulfide bonds or ester groups [27].

Wang and coworkers chemically crosslinked low molecular weight HPEI by 30-dithiobispropanoic acid (DTPA) to prepare bioreducible HPEI (SS-HPEI) [26, 28]. It is stated that low molecular weight HPEI crosslinked with disulfide containing agents have enhanced transmission efficiency than the original low molecular weight HPEI [29]. The SS-PEI complexes were used for the transfer of human telomerase reverse transcriptase (hTERT) siRNA. The in vitro and in vivo results revealed that the complexes of SS-PEI/siRNA were formed by condensation of the siRNA by the polymer matrix and was able to transfect HepG2 cells efficiently by condensing the siRNA into compact complexes. It also displayed low cytotoxicity due to the easy bond cleavage of SS-HPEI (Fig. 5.6). The results also showed that the complexes of SS-PEI/siRNA obstructs the growth of HepG2 tumor with no unfavorable effect on liver or kidney in a xenograft mouse model.

Hyperbranched poly(ester-amine)s (HPEAs) have been considered as a substitute for hyperbranched polyamines. It exhibits low cytotoxicity with efficient gene transfection at the specified site. They contain hydrolytically degradable ester groups in the structure which makes it readily biodegradable materials. HPEAs

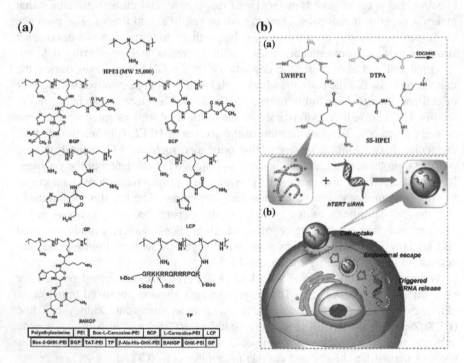

Fig. 5.6 a Graphical representation of HPEI; **b** schematic representation of a synthesis strategy of bioreducible SS-HPEI, b siRNA delivery by SS-HPEI. Reprinted with permission from Refs. [26, 28]. Copyright (2011, 2012) Elsevier

possess the characteristics similar to polyamines due to the presence high density of primary amines at the terminal ends, along with that they exhibit excellent biodegradability. Thus, HPEAs are considered as a promising gene delivery vehicle because of their controlled release capability and minimal cytotoxicity, along with a significant structural diverseness.

Different groups have tried to study about the transfection efficiency of HPEA by using different kinds of amine in the structure. For example, Liu and coworkers polymerized trifunctional amine monomers 1-(2-aminoethyl)piperazine with 1,4-butanediol diacrylate via Michael addition polymerization. The biodegradable HPEA prepared thus possessed primary, secondary and tertiary amines simultaneously [30]. The different types of amine groups affected the condensation of DNA as well as increased the ability to promote escape of vectors from lysosomes by enhancing the pH-buffering capacity. Even with high polymer/DNA weight ratio (approx 30:1), the HPEA vehicle exhibited negligible cytotoxicity and high transference ability.

Similarly, Park and coworkers reported synthesis of cationic HPEA with a hybrid structure of biodegradable ester backbone, primary amine present at the periphery, and tertiary amine groups present in the interior [31]. This biodegradable cationic polymer showed minimum toxicity and could condense negatively charged DNA.

Further Mikos and coworkers prepared a series of HPEAs keeping the amine monomer, 1-(2-aminoethyl)piperazine, same in each case and modified the structures by using different types of triacrylate monomers as spacer between the polymeric chains. They varied the structure of the spacer as they tried to analyze the relation between hydrophilic spacer lengths and HPEA properties related to the gene transference process [32]. The results revealed that in the presence of hydrophilic spacers in the structure of HPEAs the cytotoxicity of the resultant polymer decreased along with the increment in hydrolytic degradation rate. In addition to that the variation in the spacer structure also decreased the reduced charge density of HPEAs, which in turn influenced the end properties of the polymer such as polymer stability, DNA condensation, and enhanced endosomal escape.

Feijen and coworkers tried to develop a water-soluble, degradable gene carriers with very low cytotoxicity. They designed a HPEAs-based gene delivery vehicle consisting of primary, secondary and tertiary amino groups in the structure. The resultant HPEAs were capable of condensing plasmid DNA into nanostructured positively charged complexes [33]. From the results it was concluded that these water-soluble HPEAs had much better transfection efficiency than that of PEI and PDMAEMA.

As a substitute for polyamines, various hyperbranched poly (amidoamine) (HPAMAM)-based hyperbranched polymers have been considered for gene transfer. Though it has similar structure to that of HPEI, but it has more advantages such as peptide-mimicking properties, reduced haemolytic activity, excellent biocompatibility/biodegradability, and low toxicity. Generally three kinds of parameter is considered to study the efficiency of HPAMAM-based gene delivery vehicles. First parameter is the branching architecture of the hyperbranched polymer. Increasing branches in the structure greatly makes the polymer more compact with high density of terminal functional groups. Thus the gene transfection capability also gets affected by the increasing degree of branching [34]. This concept was studied by Zhu and his coworkers. They synthesized a series of cationic HPAMAMs through surface grafting polymerization of *N*,*N*-methylenebisacrylamide and 1-(2-aminoethyl)piperazine monomers. Though they had similar compositions and molecular weights, but a variation of degree of branching was done to generate different set of copolymers [35]. It was found that with increasing DB, the DNA condensation capability of HPAMAM was improved significantly along with decreased cytotoxicity (Fig. 5.7). The observation could be explained on the basis that with increasing branching, the structure became more compact and there was more availability of primary and tertiary amino groups. Consequently, the efficiency of gene transfection was enhanced on a large scale.

Secondly, the gene transfection activity of HPAMAM can be altered by introducing various functionalities at the periphery of the polymer. The alteration of terminal groups and its direct effect on transfection efficiency has been examined by Gao and coworkers [36]. They modified HPAMAM by attaching phenylalanine to the terminal amine groups of the polymer and was able to improve the bioactivity of the polymer. Third, another method to enhance transfection ability and reduce cytotoxicity in HPAMAM is to attaché biodegradable linkages introduction of

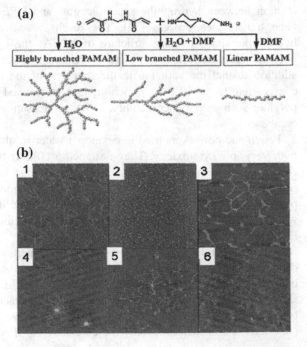

Fig. 5.7 a Schematic route for the synthesis of different structural PAMAMs (from highly branched to linear), **b** AFM images of DNA condensation by various branched PAMAMs with varying DB (0.44, 0.31, 0.21, 0.11, and 0.04, respectively). Reprinted with permission from Ref. [35]. Copyright (2010) American Chemical Society

biodegradable linkages (such as reducible disulfide bonds or ester groups) in the backbone of the base polymer.

Zhou and coworkers prepared a series of amphiphilic hyperbranched copolymer with a hydrophobic PEHO (poly (3-ethyl-3-(hydroxymethyl)-oxetane) cores and PDMAEMA arms with variable length for gene transfection [37]. The PEHO core had variable degree of branching. They wanted to study structure property relationship with the gene transfection efficiency. The results revealed that this copolymer, PEHO-g-PDMAEMAs, much better transfection efficiency than HPEI and PDMAEMA homopolymers and the efficiency was enhanced with the increasing branching in PEHO cores (Fig. 5.8). The HBs has very low cytotoxicity also which makes them a potential gene delivery vehicle.

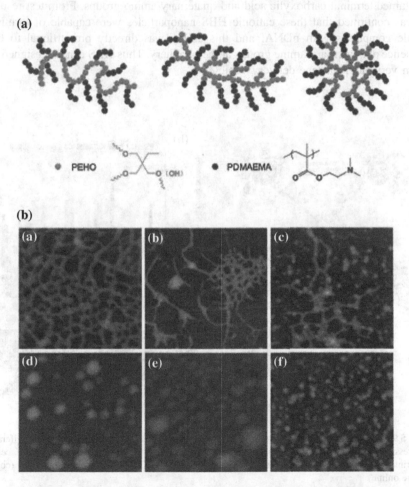

Fig. 5.8 a Schematic illustration of different architectures of PEHO-g PDMAEMA copolymers with varying DB. b AFM images of copolymer—pDNA complexes Reprinted with permission from Ref. [37]. Copyright (2012) Royal Society of Chemistry

Zhu and coworkers studied about the primary- or tertiary amine-modified β-CD derivatives and an AD-modified HPG for gene delivery (Fig. 5.9). The amine-modified β-CD compromised of per-6-amino-β-CD with seven primary amines and per-6-dimethylaminoethyl-β-CD with seven tertiary amines. The gene transfection efficiency of these charge-tunable supramolecules was based on the conventional host–guest interactions. The surface charge density of the resultant HPEAs was controlled by manipulating the molar ratios of the two cationic β-CD derivatives. These poly(etheramine)s can be considered as an efficient gene delivery vehicle with very low cytotoxicity and enhanced transfection efficiency than the standard HPEI [38].

Lee and coworkers reported poly(etheramine)s based on hyperbranched polysiloxysilane (HBS) and studied its gene transfection ability [39]. The HBS contained terminal carboxylic acid and quaternary amino groups. From test results it was confirmed that these cationic HBS nanoparticles were capable of forming stable complexes with pDNA, and this ability was directly proportional to the presence of quaternary amine groups at the periphery. Thus HBS can be designed to form very efficient gene delivery vehicle.

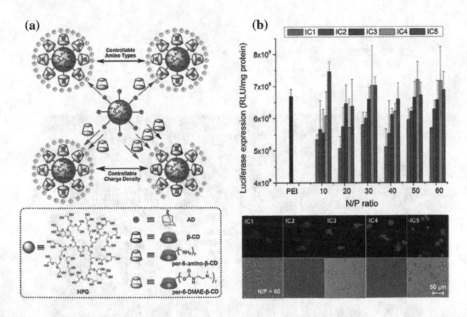

Fig. 5.9 a Schematic illustration of the synthesis of the charge-tunable HPEAs **b** Luciferase expression (*top*) and *green* fluorescent protein expression (*bottom*) of HPEAs in COS-7 cells. Reprinted with permission from Ref. [38]. Copyright (2011) Royal Society of Chemistry (color figure online)

5.6 Conclusion

In the last few years, a significant progress in the field of biocompatible or biodegradable hyperbranched polymers has been observed, an important subclass of architectural macromolecules. They can be easily modified to achieved tailor-made properties for specialized purposes. Due to their unique topological structures, abundant functional groups, low toxicity, non-immunogenicity, hyperbranched polymers prove to be of great potential for controlled release of therapeutic agents. Biocompatible or biodegradable assemblies formed by the hyperbranched polymers exhibits excellent delivery efficiency. This field needs to be further researched for more better and specific targeting as well as for more better control over the release kinetics by the hyperbranched vehicles.

References

1. Lee CC, MacKay JA, Fr'echet JMJ, Szoka FC (2005) Nat Biotechnol 23:1517
2. Yang H, Kao WJ (2006) J Biomater Sci Polymer 17:3
3. Radowski MR, Shukla A, von Berlepsch H, Bottcher C, Pickaert G, Rehage H, Haag R (2007) Angew Chem Int Ed 46:1265–1269
4. Zou J, Shi W, Wang J, Bo J (2005) Macromol Biosci 5:662–668
5. Ye L, Letchford K, Heller M, Liggins R, Guan D, Kizhakkedathu JN, Brooks DE, Jackson JK, Burt HM (2010) Biomacromol 12:145–155
6. Kainthan RK, Mugabe C, Burt HM, Brooks DE (2008) Biomacromol 9:886–895
7. Kra¨mer M, Kopaczynska M, Krause S, Haag R (2007) J Polym Sci Part A: Polym Chem 45:2287–2303
8. Liu JY, Pang Y, Huang W, Zhu XY, Zhou YF, Yan DY (2010) Biomaterials 31:1334–1341
9. Chen S, Zhang XZ, Cheng SX, Zhuo RX, Gu ZW (2008) Biomacromol 9:2578–2585
10. Liu CH, Gao C, Yan DY (2006) Macromolecules 39:8102–8111
11. Perumal O, Khandare J, Kolhe P, Kannan S, Lieh-Lai M, Kannan R (2009) Bioconjugate Chem 20:842–846
12. Lee S, Saito K, Lee HR, Lee MJ, Shibasaki Y, Oishi Y, Kim BS (2012) Biomacromol 13:1190–1196
13. Zhuang YY, Su Y, Peng Y, Wang DL, Deng HP, Xi XD, Zhu XY, Lu YF (2014) Biomacromol 15:1408–1418
14. Haxton KJ, Burt HM (2008) Dalton Trans 43:5872–5875
15. Prabaharan M, Grailer JJ, Pilla S, Steeber DA, Gong S (2009) Biomaterials 30:5757–5766
16. Chen C, Liu G, Liu X, Pang S, Zhu C, Lv L, Ji J (2011) Polym Chem 2:1389–1397
17. Tai H, Tochwin A, Wang W (2013) J Polym Sci A Polym 51:3751–3761
18. Graul AI, Lupone B, Cruces E, Stringer M (2013) Drugs Today 49:33–68
19. Salmaso S, Caliceti P (2013) Int J Pharm 440:111–123
20. Jiang M, Wu Y, Heb Y, Nie J (2009) Polym Int 58:31–39
21. Wurm F, Klos J, Räder HJ, Frey H (2009) J Am Chem Soc 131:7954–7955
22. Gao X, Zhang X, Wu XZ, Zhang X, Wang Z, Li C (2009) J Controlled Release 140:141–147
23. Zhang X, Zhang X, Wu Z, Gao X, Shu S, Wang Z, Li C (2011) Carbohydr Polym 84:1419–1425
24. Verma IM, Somia N (1997) Nature 389:239–242
25. Kadlecova Z, Rajendra Y, Matasci M, Baldi L, Hacker DL, Wurm FM, Klok HA (2013) J Controlled Release 169:276–288

26. Dey D, Inayathullah M, Lee AS, LeMieux MC, Zhang X, Wu Y, Nag D, De Almeida PE, Han L, Rajadas J, Wu JC (2011) Biomaterials 32:4647–4658
27. Jiang HL, Kim YK, Arote R, Nah JW, Cho MH, Choi YJ, Akaike T, Cho CS (2007) J Controlled Release 117:273–280
28. Xia W, Wang P, Lin C, Li Z, Gao X, Wang G, Zhao X (2012) J Controlled Release 157:427–436
29. Gosselin MA, Guo W, Lee RJ (2001) Bioconjugate Chem 12:989–994
30. Liu Y, Wu D, Ma Y, Tang G, Wang S, He C, Chung T, Goh S (2003) Chem Commun 2630–2631
31. Lim Y, Kim SM, Lee Y, Lee W, Yang T, Lee M, Suh H, Park J (2001) J Am Chem Soc 123:2460–2461
32. Chew SA, Hacker MC, Saraf A, Raphael RM, Kasper FK, Mikos AG (2009) Biomacromol 10:2436–2445
33. Zhong Z, Song Y, Engbersen JFJ, Lok MC, Hennink WE, Feijen J (2005) J Controlled Release 109:317–329
34. Zhang B, Ma X, Murdoch W, Radosz M, Shen Y (2013) Biotechnol Bioeng 110:990–998
35. Wang RB, Zhou LZ, Zhou YF, Li GL, Zhu XY, Gu HC, Jiang XL, Li HQ, Wu JL, He L, Guo XQ, Zhu BS, Yan DY (2010) Biomacromol 11:489–495
36. Wang X, He Y, Wu J, Gao C, Xu Y (2010) Biomacromol 11:245–251
37. Yu SR, Chen JJ, Dong RJ, Su Y, Ji B, Zhou YF, Zhu XY, Yan DY (2012) Polym Chem 3:3324–3329
38. Dong RJ, Zhou LZ, Wu JL, Tu CL, Su Y, Zhu BS, Gu HC, Yan DY, Zhu XY (2011) Chem Commun 47:5473–5475
39. Kim WJ, Bonoiu AC, Hayakawa T, Xia C, Kakimoto M, Pudavar HE, Lee KS, Prasada PN (2009) Int J Pharm 376:141–152

Chapter 6
Part II: In Bioimaging

6.1 Introduction to Diagnosis via Bioimaging

Medical imaging is the process of creating visual images of the interior of a living body for clinical analysis and medical treatment, as well as to provide visual representation of the function of some organs or tissues. Medical imaging is done to reveal internal structures hidden by the skin and bones, so as to diagnose the disease accurately and treat it properly. Through medical imaging, a database of normal anatomy and physiology is also established to make it possible to mark the abnormalities. Bioimaging is the emerging field of research where a sophisticated bioimaging probe is combined with modern advanced modules to reveal the specific molecular pathways in vivo [1]. For locating abnormalities and to detect various biological diseases, the most common bioimaging techniques that are used nowadays are optical imaging, MRI, single-photon emission computed tomography (SPECT), positron emission tomography (PET), and others [2]. These modules give good results but still there are some limitations that restrict their clinical applications. Nonspecificity, toxicity, and instability are the major drawbacks with these biological probes, along with that there is lack of functionality as well [3]. Therefore, a wide range of bioimaging probes have been invented till date with much enhanced detection sensitivity and selectivity but compared to small-molecule probes, polymer-based bioimaging probes have gained more interest of researchers due to its several unique characteristics. Polymer-based bioimaging probes have better stability, low cytotoxicity, specific targeting ability, and prolonged plasma half-lives. Again among different architectural polymers, hyperbranched polymeric probes are more preferred due to its easy one-pot synthesis technique, along with an alterable highly branched architecture with a wide range of functional groups present at the terminal end.

As already discussed in the previous chapters, hyperbranched polymers have a great encapsulation ability, so they can easily entrap the imaging probes by covalent or noncovalent interactions. Generally, hyperbranched polymers are combined with

© Springer Nature Singapore Pte Ltd. 2018
A. Bandyopadhyay et al., *Hyperbranched Polymers for Biomedical Applications*,
Springer Series on Polymer and Composite Materials,
https://doi.org/10.1007/978-981-10-6514-9_6

Fig. 6.1 Hyperbranched
polymer-based bioimaging
probes. Reprinted with
permission from Ref. [2].
Copyright (2013) Royal
Society of Chemistry

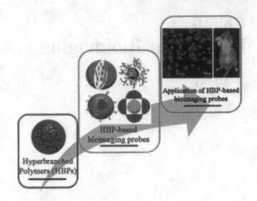

various imaging modules to produce different bioimaging probes as for optical
imaging, MRI, nuclear imaging, and ultrasound imaging. The main advantage with
the hyperbranched polymers based on bioimaging probes is that the complex
biological phenomenon can be visualized at the molecular level [2]. Herein, this
chapter discusses the remarkable advances in hyperbranched polymeric probes for
bioimaging and diagnosis (Fig. 6.1).

6.2   Hyperbranched Polymers as Fluorescent Probes

The highly sensitive optical imaging is the most common tool used for bioimaging
and is much preferred over MRI or PET due to low cost. But the major drawbacks
of optical imaging are light scattering, autofluorescence, and absorption by biological tissues which is overcome by combining it with hyperbranched polymers
[4]. The combination of hyperbranched polymers and optical bioimaging has lead
to the generation of hyperbranched polymer-based fluorescent probes, which has a
great potential for clinical implementations. Hyperbranched polymer-based
fluorescent probes can be of two types—either hyperbranched polymers form
complexes with fluorescent materials such as small fluorophores, fluorescent proteins, and inorganic fluorescent agents, or the hyperbranched polymers possess
fluorescence properties itself.

Though organic fluorescent dyes are widely used as fluorescent components of
bioimaging probes, there are certain disadvantages with these components—low
photobleaching thresholds, short half-life in blood, poor membrane permeability to
live cells, and the lack of specificity for their target cells, tissues, or organs [5]. To
solve these problems, HB-conjugated or HB-encapsulated organic fluorescent
agents are being developed.

Fluorescein isothiocyanate (FITC) is one of the most used fluorescent dyes.
FITC was easily covalently linked to HBPs for bioimaging. Hyperbranched poly
(sulfone-amine) (HPSA) was prepared and labeled with FITC by combining the
isothiocyanate group of FITC and amino group of HPSA [6]. HPSA exhibits low

cytotoxicity and good serum compatibility, while FITC-labeled HPSA reduces endosomal degradation.

Other than FITC, DOX is another most commonly used chemotherapy drug. Due to its autofluorescence property, DOX is considered as a good candidate for bioimaging. Prabaharan and his group synthesized amphiphilic conjugation of DOX and folic acid with H40-star-[poly(L-aspartate)-b-poly(ethylene glycol)] to form a tumor-targeting drug delivery system, {folated H40-star-[(PLA-DOX)-b-PEG]}, which can be conjugated to HBPs to form an imaging probe. The fluorescence of DOX was used directly to measure the cellular uptake without additional markers. The presence of folic acid enhanced cellular internalization for the system. Confocal laser scanning microscopic images of DOX fluorescent mouse breast cancer cells (4T1) are given in Fig. 6.2 [7].

Zhou and coworkers modified HBPO-star-PEG with a targeting ligand, aptamer, and fluorescent probe carboxyfluorescein. These hyperbranched polymer conjugates showed great potential for targeted cancer imaging (Fig. 6.3) [8]. The conjugate was able to form co-assembly to generate mixed micelle. When MCF-7 cells were cultured under this medium, it showed great biocompatibility and very low cytotoxicity. Bright green fluorescence was observed in MCF-7 cells after cultured with the mixed micelles, confirming the excellent cancer cell imaging and cancer cell targeting capability of the conjugate.

Inorganic quantum dots (QDs) have also great potential for bioimaging and targeting biomarkers due to its high photo-stability, broad excitation wavelength, and narrow emission spectra [9]. Due to nanocavities present in the hyperbranched structure, HB provides a great matrix for generation of QDs.

Zhu and coworkers designed and synthesized double-hydrophilic multiarm hyperbranched polymer HPAMAM-star-PEG with dendritic HPAMAM cores and many linear PEG arms linked with pH-sensitive acylhydrazone linkages. The structure has many voids present which can be used as a nanoreactor for CdS QD synthesis (Fig. 6.4) [10]. The hyperbranched polymer provides high stability, low

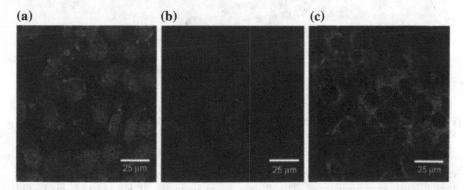

Fig. 6.2 CLSM images of 4T1 cells with **a** Free DOX, **b** H40-star-[(PLA-DOX)-b-PEG], **c** folated H40-star-[(PLA-DOX)-b-PEG]. Reprinted with permission from Ref. [7]. Copyright (2009) Elsevier

Fig. 6.3 Schematic illustration of the co-assembly of aptamer-functionalized and fluorescently functionalized hyperbranched copolymers and their cell imaging. Reproduced with permission from Ref. [8]. Copyright (2014) American Chemical Society

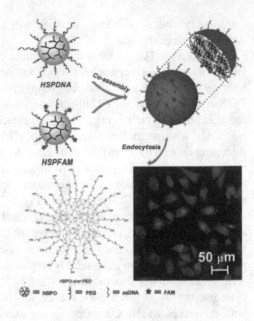

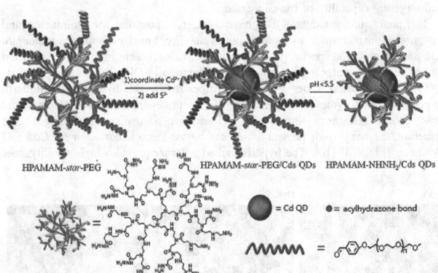

Fig. 6.4 Schematic illustration of pH-sensitive HPAMAM-star-PEG–CdS QD nanocomposites. Reproduced with permission from Ref. [10]. Copyright (2010) American Chemical Society

cytotoxicity, and pH-responsive characteristics, whereas CdS QDs provide excellent fluorescence properties. The fluorescence intensity of HPAMAM-star-PEG–CdS QD nanocomposites was controlled by adjusting the pH value. When COS-7 cells (a cell line derived from kidney cells of the African green monkey) were added into the culture medium along with these nanocomposites, due to the cleavage of acylhydrazone bonds between HPAMAM and PEG in low pH medium, the fluorescence intensity in COS-7 cells increased greatly with increasing culture time. Therefore, these pH-responsive HPAMAM-star-PEG–CdS QD nanocomposites hold great potential as novel fluorescent probes in cells.

Hyperbranched-conjugated polymers (HCPs) and hyperbranched polyamines are widely studied for optical imaging in biological systems, due to their unique properties [11]. Due to the branching in the structure, the molecular interaction in the conjugated segment is diminished. Further, hydrophilic group can be introduced in the structure for better solubility in aqueous medium.

Zhu and coworkers synthesized a unimolecular micelle from HCP-star-PEG having a HCP core and many PEG arms that can self-assemble to form multi-molecular micelle. This phenomenon can be greatly utilized for bioimaging purpose. The presence of hydrophilic PEG arms prevents intermolecular aggregation of conjugated polymeric core which in turn improves the emission ability. The emission-enhanced HCP-star-PEG micelles could be used to evaluate the cellular uptake of MCF-7 cell. The strong fluorescence was observed mainly in the cytoplasm of the cells when the cells were cultured in HCP-star-PEG for 2 h, illustrating the successful cellular imaging of multiarm HCPs [12].

By combining with benzothiadiazole unit, hyperbranched-conjugated polyelectrolytes showed enhanced blue to green fluorescence spectra, which were readily used for bioconjugation and bioimaging. This phenomenon can be explained as with the addition of benzothiadiazole group, water-soluble conjugated polyelectrolytes were formed with better bioimaging properties [13].

Hyperbranched polyamines are relatively weak in their fluorescent properties. To improve the fluorescence intensity of these hyperbranched polyamines for bioimaging, Pan and his coworkers tried to develop hyperbranched poly (amine-ester) (HypET) and compare its fluorescence properties with the linear analogue. With a large number of tertiary amine groups present in HypET, it showed strong fluorescent properties. For analyzing the cell imaging properties, a water-soluble galactopyranose-modified HypET (HypET-AlpGP) was prepared by hydrolysis of 1,2,3,4-di-o-isopropylidene-α-D-galactopyranose-grafted HypET with quantum yield = 0.10 in acidic solution (pH = 3) of THF/H$_2$O. From the cytotoxicity results, it was evident that HypET-AlpGP is very less toxic than the branched PEI used as control. As analyzed from Fig. 6.5, the PEI25k exhibits high cytotoxicity to liver carcinoma HepG2 cells with 50% cell viability at the concentration of 15 µg/mL, whereas only slight decrease of the cell viability was observed for the HypET-AlpGP at a dose as high as up to 500 µg/mL. Hence, it can be concluded that HypET-AlpGP has low cytotoxicity and displays bright cell imaging [14].

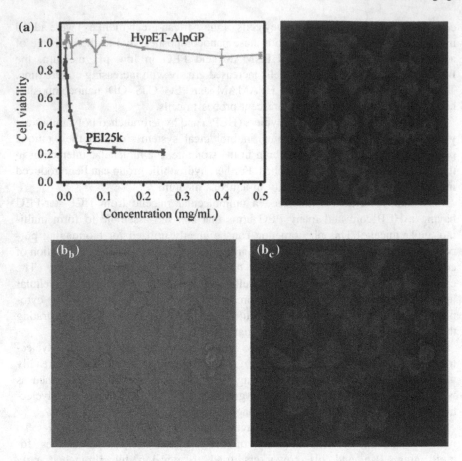

Fig. 6.5 **a** Cytotoxicity results of the HypET-AlpGP to HepG2 cells. **b** CLSM images of the HepG2 cells after 9 h incubation with a serum-free DMEM solution of HypET-AlpGP (2 mg/mL) under excitation at λex = 375 nm (ba) and bright field (bb); (bc) is a merged picture of (ba) and (bb). Reproduced with permission from Ref. [14]. Copyright (2012) American Chemical Society

6.3 Hyperbranched Polymers as MRI Contrast Agents

Magnetic resonance imaging (MRI) is a medical imaging technique used to visualize the anatomy and the physiological processes of the body. MRI scanners use strong magnetic fields, radio waves, and field gradients to generate images of the organs in the body. MRI has certain advantages over other imaging techniques such as noninvasiveness, no exposure to radiation, high spatial and temporal resolution, and excellent penetration depths toward soft [15]. But the major drawback with MRI is its low sensitivity and to overcome this issue several contrast agents have been developed which shortens the spin–lattice relaxation time (T_1) and spin–spin relaxation time (T_2) of water protons [2]. Due to low contrast efficiency, nonspecificity, and fast renal excretion of the low molecular weight contrast agents, polymeric scaffolds are preferred for attaching the MRI agents [16]. Hyperbranched polymers are more appealing due to their rigid branched structure and availability of a large number of functional along with its tunable branched architecture and size allows their selection for special purpose applications.

Generally, paramagnetic contrast agents are utilized for enhancing signal intensity as T_1 contrast agents. Sideratou and coworkers reported the synthesis of multifunctional gadolinium complexes based on commercialized hyperbranched aliphatic polyester, H40. The complex contained hydrophobic scaffold core, hydrophyllic PEG chains, ethylenediaminetetraacetic acid (EDTA), and diethylenetriaminepentaacetic acid (DTDA) as gadolinium chelates. Folic acid (FA) was also attached as targeting ligand. It was found that the HB–Gd^{3+} complexes had an increased rotational correlation lifetime when their rotation was slowed down due to the macromolecular matrix attached with the complex. Therefore, the relaxivity value of H40–EDTA–PEG–Folate and H40–DTPA–PEG–Folate gadolinium complexes was much higher than that of the conventionally used [Gd(DTPA)] complex. Furthermore, the hyperbranched complexes showed low cytotoxicity and more specificity due to FA ligands, which made them promising candidates for MRI applications [17].

Li and his coworkers tried to analyze the effect of macromolecular architecture on the relaxation phenomenon of the Gd^{3+} ligands. They took linear, star, and hyperbranched substrates. The results revealed that the linear and hyperbranched polymers have a significantly better relaxation in the ligands compared to the commercially available Gd^{3+} complexes. Though star polymer had lower relaxation than the other two grades, it was still higher than the conventional complex [18].

Superparamagnetic iron oxide nanoparticles (SPIO) have been an excellent grade of T_2-type contrast agents in MRI. Muller and his coworkers have analyzed the MR signal properties of HPG-grafted magnetic nanoparticles through immobilizing HPG onto the surface of SPIO nanoparticles via a ring-opening anionic polymerization. From the results, it was clearly evident that the HPG-g-SPIO was highly stable and capable of forming a steady dispersion in aqueous solution. The contrast imaging capability of HPG-g-SPIO solution was estimated from the spin-echo abdomen images of a living mouse examined at different intervals of time, before and after the intravenous injection applied (Fig. 6.6). A strong negative contrast was observed after 6 min from the application of the intravenous injection which prevailed in liver for 80 min and in kidney for 110 min and then gradually faded away with time. From the results, it can be concluded that HPG-g-SPIO can be considered as a stable and sensitive MRI agent [19].

Hayashi and his group tried to develop contrast agents based on organic substrates [20]. They prepared hyperbranched polystyrene containing 2,2,6,6-tetramethylpiperidine-1-oxyl (TEMPO) radical. This HPS-TEMPO-based agent exhibited better stability and relaxation than the Gd^{3+}–DTPA complexes.

Recently, ^{19}F nuclei have become an alternative approach for MRI due to its high NMR receptivity. Wooley and coworkers designed hyperbranched fluoropolymers as ^{19}F MRI agent assemblies [21]. For improving the salvation and

(a)   (b)

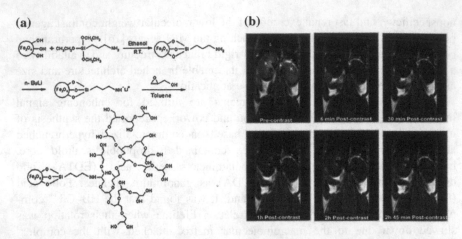

Fig. 6.6 **a** Schematic illustration of the surface modification of Fe₃O₄ nanoparticles with 3-aminopropyl triethoxysilane, and anionic polymerization of glycidol on the surface; **b** MRI images of a live mouse with a T_2-weighted spin-echo sequence at different time intervals (*L*, liver; *K*, kidney). Reproduced with permission from Ref. [19]. Copyright (2012) Wiley

mobility of the fluorinated groups, the fluorinated component was incorporated in the shell domain of hyperbranched copolymers. 4-chloromethyl styrene and lauryl acrylate were copolymerised. Trifluoroethyl methacrylate and tert-butyl acrylate were incorporated in different feed ratios to generate amphiphilic hyperbranched fluoropolymer. These HBs were capable of self-assembling into micelles with a narrow, single-resonance ^{19}F NMR signal and good T1/T2 relaxation time. These hyperbranched fluoropolymer micelles with good signal-to-noise (S/N) ratios could be used as imaging agents for ^{19}F MRI in various biomedical studies.

6.4   Hyperbranched Polymers in Nuclear Tomographic Imaging

Nuclear imaging technique is different from MRI and optical imaging as it is a quantitative measure based on gamma rays with energies. It is therefore very crucial for nuclear imaging agents to be highly sensitive toward distinguishing between primary (true) and Compton scattered (false) events. Nuclear tomographic imaging techniques are highly preferred as clinical imaging tool due to of its high accuracy, noninvasiveness, and deep penetration potency [2]. The most conventional modules used nowadays for nuclear tomographic imaging are PET and SPECT [22]. Polymers are more preferred as nuclear tomographic imaging agent due to their better specificity, low cytotoxicity, along with easy modification pathway. Again among the conventional polymers, HBs with a highly branching architecture and plentiful functional end groups offer an ideal matrix platform for radioisotopes to be utilized for nuclear imaging.

Zhu and coworkers designed hyperbranched poly(sulfone-amine) (HPSA) grafted with monoclonal antibody CH12 through N-hydroxy succinimidyl S-acetyl mercaptoacetyl triglycinate (NHS-MAG3) for labeling [188]Re, forming (CH12-HPSA-[188]Re), which could be used for the tumor detection and targeted radioimmunotherapy [23]. The in vitro results as obtained from SPECT revealed that CH12-HPSA-[188]Re could effectively bind at the targeted hepatocarcinoma tumor tissues with overexpressed epidermal growth factor receptor vIII (EGFRvIII). In comparison with free radionuclide [188]Re, CH12-HPSA-[188]Re exhibited longer circulation time in blood. The SPECT images successfully revealed the gradual accumulation of CH12-tethered HPSA-based radiopharmaceutical at the tumor site of tumor-bearing mice (Fig. 6.7).

Highly sensitive PET offers the more accurate images than SPECT but the short half-lives of PET radionuclides (e.g., ^{11}C, ^{18}F, and ^{64}Cu) limit its efficiency. To solve this problem, Gong and coworkers reported synthesis of HB labeled radionuclide. A multifunctional unimolecular micelle self-assembled from a hyperbranched amphiphilic block copolymer H40-star-[poly(L-glutamate hydrazone-doxorubicin)-b-poly(ethyleneglycol)] (H40-star-P(LG-Hyd-DOX)-b-PEG) was conjugated with the cyclo(Arg–Gly–Asp–D-Phe–Cys) peptide (cRGD, for integrin avb3 targeting) and the macrocyclic chelator (1,4,7-triazacyclononane-N,N0,N00-triacetic acid [NOTA]) for cancer-targeted PET/CT imaging and drug delivery in tumor-bearing mice [24]. The anticancer drug DOX is covalently attached with the hydrophobic part of the copolymer through pH-sensitive hydrazone linkages. Complexation of ^{64}Cu onto the resulting HBP via NOTA was beneficial for reducing the copper binding with plasma protein and thus

Fig. 6.7 Biodistribution of CH12-HPSA-[188]Re at different time intervals after tail vein injection as acquired from SPECT. Reprinted with permission from Ref. [23]. Copyright (2013) Springer

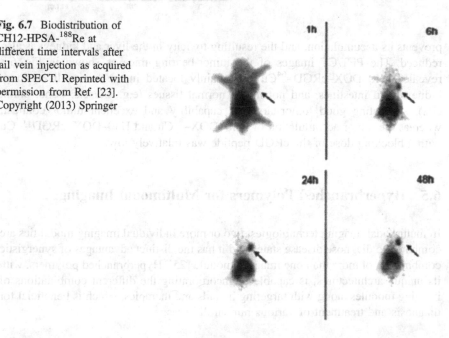

Fig. 6.8 PET/CT imaging of 64Cu-labeled nanocarriers in U87MG tumor-bearing mice. **a** Serial coronal PET images of U87MG tumor-bearing mice at various time points post-injection of H40–DOX–^{64}Cu, H40–DOX–cRGD–^{64}Cu, or H40–DOX–cRGD–^{64}Cu with a blocking dose of cRGD. **b** Representative PET/CT images of an U87MG tumor-bearing mouse at 4 h post-injection of H40–DOX–cRGD–^{64}Cu. Reproduced with permission from Ref. [24]. Copyright (2012) Elsevier

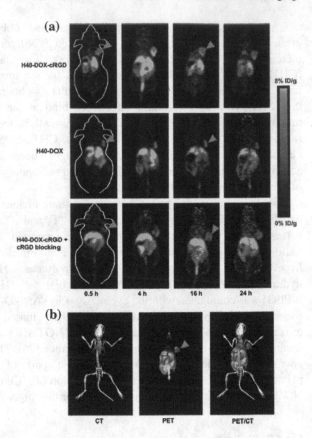

prevents its accumulation, and the resulting toxicity in the liver and kidneys is also reduced. The PET/CT images of the tumor-bearing mice as shown in Fig. 6.8 revealed H40–DOX–cRGD–^{64}Cu gets mainly located in the tumor, liver, lung, kidneys, and intestines, and not in the normal tissues (e.g., muscle, bone, brain, etc.), suggesting good tumor-targeting capability and excellent tumor contrast, whereas the tumor accumulation of H40–DOX–^{64}Cu and H40–DOX–cRGD–^{64}Cu with a blocking dose of the cRGD peptide was relatively low.

6.5 Hyperbranched Polymers for Multimodal Imaging

In multimodal imaging technologies, two or more individual imaging modalities are combined to diagnose disease states and it has the distinct advantages of synergistic combination of more than one imaging module [25]. Hyperbranched polymers, with its unique architectures, is capable of incorporating the different combinations of imaging modules along with targeting ligands and therapies, which is beneficial for diagnosis and treatment of various human diseases.

For example, Thurecht and coworkers developed multimodal molecular imaging agents based on hyperbranched polymers, which included both optical fluorescence imaging and PET/CT [26]. Two different sets of hydrophilic hyperbranched polymers with different degrees of branching were synthesized via controlled RAFT polymerization. Then the polymers were conjugated with the infrared dyes to be utilized for optical imaging along with a copper chelator capable of binding of ^{64}Cu as a PET radio nucleus. The multimodal imaging results of murine B16 melanoma model of mice revealed that the larger HB with high DB had a considerably longer circulation time and showed enhanced accumulation ability in the targeted solid tumors within the mice. The advantages of the multimodal imaging agents are evident from the fact that in this case PET module exhibited high sensitivity immediately after injection of the imaging agent, whereas the optical imaging module promotes extended longitudinal studies.

MRI images have much higher resolution while optical images are more sensitive. Clubbing these two, a new bioimaging module was constructed complementing each other. Parallely, researchers are trying to develop a combined fluorescent probe with MRI on the hyperbranched polymeric matrix. Whittaker and his coworkers tried to develop this unique multimodal bioimaging agent with synergistic effect of both the modules. They designed multimodal hyperbranched polymeric nanoparticles by attaching high-resolution ^{19}F MRI and sensitive fluorescence imaging together [27]. They prepared HBs with controlled molecular size and structure. Folic acid was attached at the terminal ends by standard coupling reactions for targeting. They used rhodamine B isothiocyanate as the fluorescent label. A mouse subcutaneous tumor model was established to investigate the effectiveness of the HB nanoparticles for molecular imaging. Hydrophilic PEG-based macrostructure was used for encapsulation of ^{19}F nuclei which exhibited segmental mobility both in serum and intracellular fluids. The results demonstrated, in the multimodal imaging probe, the total tracking of the nanoparticles could be done via fluorescence imaging, whereas ^{19}F MRI provided high-resolution images which were used for the exact estimation of the in vivo distribution of nanoparticles within single organs. As shown in Fig. 6.9, the non-targeted nanoparticles are excreted from the body as it gets accumulated mainly in the bladder and kidneys, but the FA-targeted polymeric nanoparticles got accumulated in the tumor and liver besides the kidneys and the bladder. Even the fluorescence imaging established the observations stated above as the signal from the FA-conjugated nanoparticles was recorded in the liver, bladder, kidneys, and tumor. This bimodal system uses the sensitivity and low-cost advantages of optical imaging along with the high-resolution capabilities of MRI.

The incorporation of three kinds of imaging modalities on a single imaging probe was studied by Häfeli and coworkers. They functionalized a high molecular weight HPGs with a suitable ligand for ^{111}In radiolabeling, Gd coordination, and also labeled it with a fluorescent dye. Thus, the conjugate becomes a trimodal probe utilizing the advantages of SPECT, MRI, and fluorescent imaging. The utility of combining all the three probes was evident from the in vivo results. The MR imaging of the HPG labeled with Galladium complexes provided information

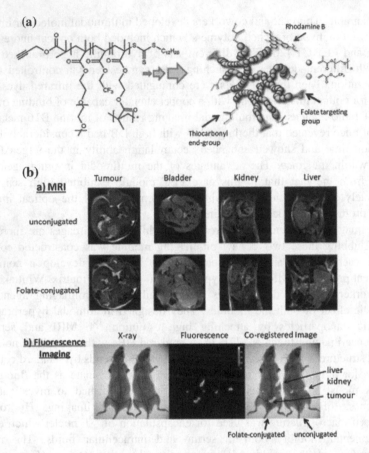

Fig. 6.9 **a** Schematic representations of the molecular structure of HBPs. **b** Molecular images of HBP nanoparticles using the mouse subcutaneous B16 melanoma model: **a** MRI images of bladder, kidneys, liver, or tumor (*circled* in image) in the tumor-bearing mice 1 h following intravenous injection of 100 mL of FA-conjugated or unconjugated (control) HBP (20 mg mL^{-1} in PBS). The high-resolution ^{1}H MR image is overlaid with the 19F image. **b** Fluorescence images of mice following injection of the same two compounds at the same concentration. Reproduced with permission from Ref. [27]. Copyright (2014) American Chemical Society (color online)

regarding vascular function, while SPECT analysis by ^{111}In-labeled HPG was able to quantitatively analyze the probe's biodistribution over time. The microregional location of the probe within the tumor microenvironment, including how long the probe stayed in intravascular environment, was evaluated at the microscopic level using the HPG tagged with fluorescent dye image optical imaging [28]. The combination of three imaging modalities exhibited great potential in preclinical investigations to provide highly specific and quantitative data regarding the physiological function of tumor blood vessels.

6.6 Conclusion

Thus it can be concluded that hyperbranched polymers with highly branched architecture prove to be an excellent agent for bioimaging applications such as optical imaging, MRI, nuclear imaging, etc.. With the presence of hyperbranched polymers, the imaging agents become more specific, more sensitive, more stable, and have longer circulation time. Hyperbranched polymers provide more stable scaffold for bioimaging probes, which makes it more interesting for biomedical researchers. In the last few years, the demand for development of various hyperbranched polymers with potential for application in bioimaging probes has increased considerably. Though several researches have been done already, still few questions are unanswered till now such as hyperbranched polymers in ultrasound imaging is still left to be explored. The accuracy of the images, proper understanding of the interaction between the hyperbranched polymer-based probe and the biological system, and improvement of therapeutic efficiency needs to be studied further. Hopefully, in the coming years, new materials would be developed with more efficiency and accuracy.

References

1. Kim JH, Park K, Nam HY, Lee S, Kim K, Kwon IC (2007) Prog Polym Sci 32:1031–1053
2. Zhu Q, Qiu F, Zhu BS, Zhu XY (2013) RSC Adv. 3:2071–2083
3. Medintz IL, Uyeda HT, Goldman ER, Mattoussi H (2005) Nat Mater 4:435–446
4. Yu Y, Feng C, Hong YN, Liu JZ, Chen SJ, Ng KM, Luo KQ, Tang BZ (2011) Adv Mater 23:3298–3302
5. Lippincott-Schwartz J, Snapp E, Kenworthy A (2001) Nat Rev Mol Cell Biol 2:444–456
6. Zhou YF, Huang W, Liu JY, Zhu XY, Yan DY (2010) Adv Mater 22:4567–4590
7. Prabaharan M, Grailer JJ, Pilla S, Steeber DA, Gong S (2009) Biomaterials 30:5757–5766
8. Yu SR, Dong RJ, Chen JJ, Chen F, Jiang W, Zhou YF, Zhu XY, Yan DY (2014) Biomacromol 15:1828–1836
9. Michalet X, Pinaud FF, Bentolila LA, Tsay JM, Doose S, Li JJ, Sundaresan G, Wu AM, Gambhir SS, Weiss S (2005) Science 307:538–544
10. Zhu LJ, Shi YF, Tu CL, Wang RB, Pang Y, Qiu F, Zhu XY, Yan DY, He L, Jin CY, Zhu BS (2010) Langmuir 26:8875–8881
11. Qiu F, Wang DL, Zhu Q, Zhu LZ, Tong GS, Lu YF, Yan DY, Zhu XY (2014) Biomacromol 15:1355–1364
12. Qiu F, Tu CL, Wang RB, Zhu LJ, Chen Y, Tong GS, Zhu BS, He L, Yan DY, Zhu XY (2011) Chem Commun 47:9678–9680
13. Pu KY, Shi JB, Cai LP, Li K, Liu B (2011) Biomacromol 12:2966–2974
14. Sun M, Hong CY, Pan CY (2012) J Am Chem Soc 134:20581–20584
15. Mulder WJM, Strijkers GJ, van Tilborg GAF, Griffioen AW, Nicolay K (2006) NMR Biomed 19:142–164
16. Langereis S, Dirksen A, Hackeng TM, van Genderen MHP, Meijer EW (2007) New J Chem 31:1152–1160
17. Sideratou Z, Tsiourvas D, Theodossiou T, Fardis M, Paleos CM (2010) Bioorg Med Chem Lett 20:4177–4181

18. Li Y, Beija M, Laurent S, Elst LV, Muller RN, Duong HT, Lowe AB, Davis TP, Boyer C (2012) Macromolecules 45:4196–4204
19. Arsalani N, Fattahi H, Laurent S, Burtea C, Vander Elst L, Muller RN (2012) Contrast Media Mol Imaging 7:185–194
20. Hayashi H, Karasawa S, Tanaka A, Odoi K, Chikama K, Kuribayashi H, Koga N (2009) Magn Reson Chem 47:201–204
21. Du WJ, Nystrom AM, Zhang L, Powell KT, Li YL, Cheng C, Wickline SA, Wooley KL (2008) Biomacromol 9:2826–2833
22. Boase NRB, Blakey I, Thurecht KJ (2012) Polym Chem 3:1384–1389
23. Li N, Jin Y, Xue L, Li P, Yan D, Zhu X (2013) Chin J Polym Sci 31:530–540
24. Xiao Y, Hong H, Javadi A, Engle JW, Xu W, Yang Y, Zhang Y, Barnhart TE, Cai W, Gong S (2012) Biomaterials 33:3071–3082
25. Thurecht KJ (2012) Macromol Chem Phys 213:2567–2572
26. Boase NRB, Blakey I, Rolfe BE, Mardon K, Thurecht KJ (2014) Polym Chem 5:4450–4458
27. Rolfe BE, Blakey I, Squires O, Peng H, Boase NR, Alexander C, Parsons PG, Boyle GM, Whittaker AK, Thurecht KJ (2014) J Am Chem Soc 136:2413–2419
28. Saatchi K, Soema P, Gelder N, Misri R, McPhee K, Baker JH, Reinsberg SA, Brooks DE, Ha¨feli UO (2012) Bioconjug Chem 23:372–381

Chapter 7
Part III: Tissue Engineering

7.1 Introduction

Hyperbranched polymers have a three-dimensional structure with high functionality, high reactivity due to the presence of a large number of free terminal groups, and they exhibit enhanced absorption capacity of biomolecules on a polymeric biomaterial. More advantage with these architectural polymers is that they can be altered structurally as well as by incorporation of functional groups can be improved for better cell attachment. Hyperbranched polymers are quite capable of forming porous hydrogels or films as scaffolds, and are promising material to support adhesion and rapid reproduction of cells. Thus, hyperbranched polymers, due to their unique structures and special properties, have proved to be of high potential in various applications in tissue engineering fields.

7.2 Hyperbranched Polymers as Tissue Scaffold Component

A wide variety of hyperbranched polymers, such as hyperbranched poly(lactic acid) (PLA), poly(lactic-glycolic acid) (PLGA), polycaprolactone (PCL), polyurethane, polyethylene glycol (PEG), polyglycerol (PG), poly(NIPAM) have been extensively examined and widely studied for fabrication of tissue engineering scaffolds [1–6].

With high number of functional end groups, biocompatible hyperbranched polymers provide a more densely packed matrix and has proved to be suitable as an ideal biological scaffold with the following characteristics: (1) it possesses a porous three-dimensional network that is resolvable in vivo, (2) it is comparable with the mechanical properties of the normal human cartilage, (3) it promotes cell growth in the surrounding joint area, (4) it does not alter the immunity system of the body,

© Springer Nature Singapore Pte Ltd. 2018
A. Bandyopadhyay et al., *Hyperbranched Polymers for Biomedical Applications*,
Springer Series on Polymer and Composite Materials,
https://doi.org/10.1007/978-981-10-6514-9_7

(5) it bears high strength which can be placed in joint area and it is able to withstand the physiological loads until the tissue repair is complete, (6) it resists excess swelling of the matrix [7].

Due to the above-mentioned advantages provided by hyperbranched polymers, it is more preferred as tissue scaffolds. First, we will consider polyurethane-based tissue scaffolds. Polyurethanes with a biodegradable linear aliphatic polyester part and as its copolymers possess high molecular weight along with high glass transition temperature and modulus. The main advantage with HPU-based biomaterials is their mechanical strength, flexibility, and chemical and biological properties. Thus, they find a wide range of application in the biomedical field—from cardiovascular repair, cartilage implant, ligament regeneration, and bone replacement to controlled drug/gene delivery but mainly considered for hard-tissue scaffolds [8].

Karak and co-workers synthesized sunflower oil-based HPUs with different weight percentages of the branching agent, pentaerythritol. It was a remarkable research work as this was the first time that vegetable oil-based HBPs were prepared and it was examined as a potential scaffold material for tissue engineering [9]. The MTT–hemolytic assay and subcutaneous implantation in Wistar rats followed by cytokine–ALP assay and histopathology studies confirmed a better biocompatibility of HPU (Fig. 7.1).

Other than HBPU, aliphatic polyesters are also regarded as a good tissue scaffold component due to its easily degradable ester backbone, which breaks down easily by hydrolysis and makes adequate space for newly developing tissues. Therefore, glycolic acid (PGA), lactic acid (PLA), and their block copolymer poly (lactic-co-glycolic acid) (PLGA) are quite preferred as medical scaffolds and implants. As we already know that a hyperbranched polymer is less crystalline and has lower solution/melt viscosity. Therefore, processing is easier for hyperbranched polymers than their linear analogues with similar molecular weight. This property was utilised by Chiellini and his group. They used a branched PCL to generate a three-dimensional mesh by simple wet spinning method [10]. They further studied these meshes for regeneration of bone tissue along with their antimicrobial property. The biocompatibility of the branched PCL and the influence of the scaffold architecture on cell behavior were examined with MC-3T3 pre-osteoblast cells.

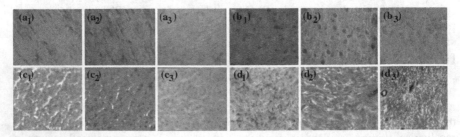

Fig. 7.1 Histological sections for heart (a_1, a_2, and a_3), for kidney (b_1, b_2, and b_3), for liver (c_1, c_2, and c_3), and for skin (d_1, d_2, and d_3) of control (1), HBPU (2), and SHBPU (3) of Wistar rats. Reprinted with permission from Ref. [9]. Copyright (2013) Wiley (Color figure online)

After 14 days of culture, from the cell adhesion and proliferation analysis it was evident that the meshes could be an alternative for the conventional materials used as engineered bone scaffolds (Fig. 7.2).

Song and his co-workers worked on amorphous shape memory polymer (SMP) network crosslinked from a star-branched macromer. It had polyhedral oligomeric silsequioxane (POSS) nanoparticle core and adjacent PLA arms [11]. The rigid POSS nanoparticle core promotes greater participation of the urethane-tethered PLA arms in the elastic deformation and the recoiling process with reduced excessive chain entanglement. Consequently, the resulting POSS-SMP nanocomposites, with cortical bone-like modulus (B2 GPa) at body temperature, could stably hold their temporary shape for more than an 1 year at room and body temperatures and achieve full shape recovery within a matter of seconds. The group further uses rat subcutaneous implantation model to examine degradation profile of these SMPs. The results showed that the degradation rates were inversely related to the length of the PLA chains within the crosslinked

Fig. 7.2 Biological characterization of PCL meshes/MC3T3-E1 constructs. **a** Live/dead visual viability assay: viable (*green* fluorescence) and dead (*red* fluorescence) cells on scaffold. Scale bar corresponding to 200 mm, applicable to both micrographs. **b** Cytochemical analysis of the constructs: toluidine *Blue* staining shows the presence of cells on scaffold and their localization along fibers. Scale bar corresponding to 100 mm, applicable to both micrographs. **c** CLSM analysis of the constructs: actin filaments (*green*) and nuclei (*blue*) showing cellular morphology and distribution on scaffold fibers. Scale bar corresponding to 200 mm, applicable to both micrographs. Reprinted with permission from Ref. [10]. Copyright (2011) Sage publishers

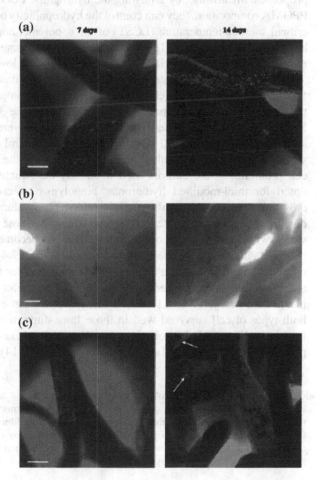

amorphous network. After 1 year of implanting these SMPs, no significant no pathologic abnormities were detected from the organs examined which makes the material ideal as a tissue scaffold material.

Hyperbranched PEGs are also considered as a potential scaffold for cell encapsulation and culture. Through various living and controlled radical polymerisation, a wide variety of hyperbranched PEGs have been developed with more controlled structure, well-defined chain lengths and with different extents of degree of branching. Lutz and his co-workers tried to modify linear poly (MEO_2MA-co-OEGMA) by introducing multifunctional vinyl monomer ethylene glycol dimethacrylate (EGDMA) within the structure [12]. They wanted to develop hyperbranched architecture but failed to do it due to microgelation formation. This method was later upgraded by Tai and his groups and they synthesized hyperbranched PEGMEMA–PPGMA–EGDMA via a one-step deactivation enhanced ATRP approach. The introduction of the multi-vinyl crosslinker EGDMA enables the copolymer with the capability of enhanced and tailorable photo-crosslinkable properties. Meanwhile, by adjusting the hydrophilic PEGMEMA and hydrophobic PPGMA composition, they can control the hydrophilicity of the polymer as the lower critical solution temperature (LCST) of the copolymer could be maintained around human body temperature [13]. The toxicity of these materials were examined over mouse C2C12 myoblast cells and they displayed very low cytotoxicity (Fig. 7.3).

Wang and his group tried to develop tissue scaffold component by varying the "long" and "short" PEG chain monomer composition and thus tried to maintain the LCST value of the copolymers around 37 °C. The 3T3 mouse fibroblast cell line was encapsulated in the hydrogel and the results were positive as there was no significant difference of cell viability between the control (cells alone) and polymer samples after 4 days of incubation. Further to improve the cytocompatibility and the cell proliferation, the authors tried to modify the structure of the polymer. They opted for thiol-modified hyaluronan biopolymer as crosslinker instead of UV crosslinking systems because of clinic safety. The hyaluronan crosslinked polymer displayed porous semi interpenetrating structure whose pore sizes and porosity varies with the polymer concentration. Thus, it becomes easier to optimize the hydrogel efficiency as tissue scaffold component (Fig. 7.4). 3T3 fibroblast cells and rabbit adipose-derived stem cells (ADSCs) were used for further biomedical studies and the results demonstrated the good cell viability after the cells were embedded inside the hydrogel. The Live/Dead assay showed that even after 1-week culture, both types of cell survived well in those three-dimensional hydrogels. The group further studied the behavior of encapsulated ADSCs and identified the secretion profile of suitable growth factors for wound healing [14].

◄Fig. 7.3 Light phase control microscope images for the cells cultured **a** in PEGMEMA-PPGMA-EGDMA copolymer culture media solutions (750 μg/mL); **b** in the culture media without polymers, **c** on the photo-cross-linked polymer films. **d** Live/Dead viability assay for the cells cultured in the copolymer/culture media solutions after 5 days. The viable cells *fluoresce green*, whereas the nonviable cells fluoresce red (pointed by the *arrow*). Reprinted with permission from Ref. [13]. Copyright (2009) American Chemical Society (Color figure online)

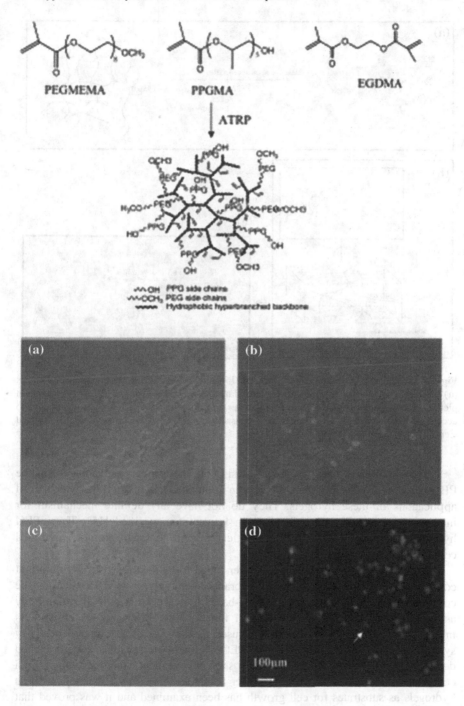

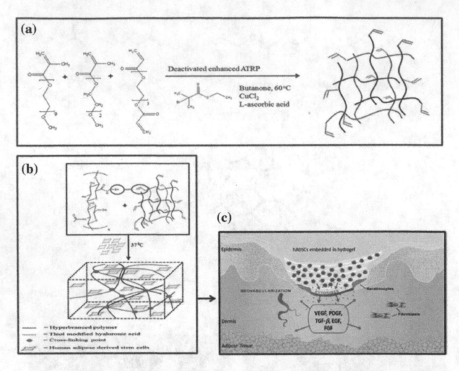

Fig. 7.4 Schematic illustration of PEGMEMA475–MEO2MA–PEGDA258 copolymer synthesis (**a**) and encapsulation of hADSCs in the crosslinked P–SH–HA hydrogel (**b**) and the cartoon picture of application of P–SH–HA hydrogel on a skin wound (**c**) for secretion of growth factors to accelerate wound healing. Reprinted with permission from Ref. [14]. Copyright (2013) Biomed Central

Large flexibility during material design is the primary advantage for the PEG-based hydrogels, but there are certain limitations that restricts the wide spread applications of these hyrogels. They do not have any definite mechanism for interacting with cells and cell adhesion is also non-specified [15]. Thus, PEG hydrogels are often modified using peptides or phosphates to promote improved cellular interactions [16, 17].

Another important subclass of polymers need to be mentioned as an important component for tissue scaffolds is hyperbranched polyglycerols (HPGs), it can be considered as a good substitute for PEG-based hydrogels due to its hydrophilicity as well as the high extent of hydroxyl functionality. Frey and his co-workers initiated the concept that HPG can be used as tissue scaffold component. They synthesized HPG hydrogels based on PEO multiarm stars with a hyperbranched dendritic core. The hydrogel products showed excellent stability with a high compression module due to its polyether backbone [18]. The efficiency of these hydrogels as substrates for cell growth has been examined and it was proved that they hold a good potential for cell growth and tissue engineering purpose.

Later, the biocompatibility of HPGs was examined by Brooks and his group using fibroblast and endothelial cells (Fig. 7.5) [19]. The assay results showed remarkably low cytotoxicity of HPG against both the cell lines.

Derivatives of HPG were also studied for tissue engineering applications. The structure was modified by attaching hydrophobic C18 alkyl chains as well as PEG-350 chains to a certain fraction of the polyether polyol OH groups [20]. Due to high solubility and low viscosity of these hyperbranched polymers, along with easy synthetic approach, they are considered as human serum albumin (HSA) substitutes. They provide a greater advantage over native HSA where there is risk of transmission of diseases. Plasma half-lives as high as 34 h as well no risk of any disease transmission makes these modified HPG suitable as synthetic plasma expanders (Fig. 7.6).

Hennink and co-workers developed dimensionally stable networks from HPG by modifying the end hydroxyl group of the polymer HPG with the photo-crosslinkable acrylate. The HPGs showed low swelling capability—thus the three-dimensional networks were highly stable. A number of sets were prepared by varying the degree of substitution and the properties of the HPG could be altered by

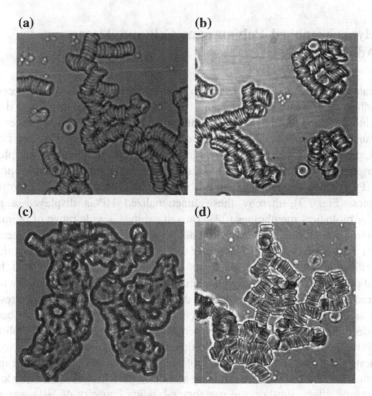

Fig. 7.5 Images of human blood red cells in anticoagulated plasma after 2 h incubation with **a** HPG, **b** LPG, **c** hetastarch, and **d** saline. Reprinted with permission from Ref. [19]. Copyright (2006) American Chemical Society

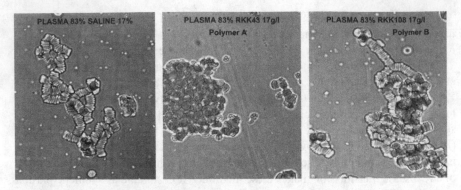

Fig. 7.6 Effect of dHPG polymers on red cells suspended in partially diluted plasma. Reprinted with permission from Ref. [20]. Copyright (2008) Elsevier

varying the concentration of HPG–MA in the aqueous solution as well as by the degree of substitution [21].

7.3   Hyperbranched Polymers as Cell and Tissue Adhesives

Hyperbranched polymers have proved to hold an enormous potential for cell and tissue adhesive. Researchers all around the world have been trying to develop various biomedical agents by utilizing the multiple functionalization properties of hyperbranched polymers. The most significant work has been done by Brooks, Kizhakkedathu, and colleagues. They synthesized HPGs containing multiple choline phosphate (CP) groups. CP has the exact inverse orientation of phosphatidyl choline (PC), which is the end group of the major lipid presented in eukaryotic cell membranes (Fig. 7.7), thereby these functionalized HPGs displayed a strong affinity for biological membranes [22]. Thus, it exhibits a wide range of biomedical application—from tissue sealing to drug delivery. In particular, the researchers took two HPG samples of different molecular weights (23 and 65 kDa) and varied the concentration of CPs attached to them. The multivalent CP-terminated hyperbranched polymers showed strong attachment with human red blood cells, where as the PC-modified polyglycerols bound to the cells very weakly. From fluorescence labeling and tritiation results, it was concluded that the strong interaction between the CP-decorated hyperbranched polymers and the PC-terminated phospholipids was mainly due to the formation of multiple CP–PC heterodimers by electrostatic interactions. Also, a comparison of molecular weights revealed an unexpected result—that for 23 kDa HPG, the binding was stronger which could be due to more stable entropy effect. Further, the membrane-binding capacity of HBs was examined using Chinese hamster ovary cells, where the authors found that the CP-terminated HPGs were rapidly taken up by the cells unlike the PC-modified

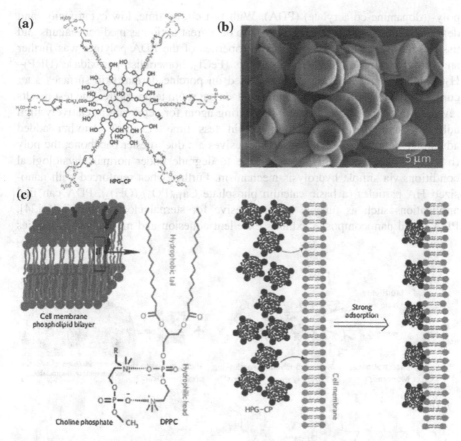

Fig. 7.7 Chemical structure of HPG–CP: multivalent HPG structures (*black*) with CP end groups (*red*) linked by 1,2,3-triazol units (*green*) (**a**) and SEM images (5000) of red blood cells forming aggregates in saline solution as a result of the cell adhesion (**b**) and the mechanism of the biomembrane adhesion interaction (**c**). Reprinted with permission from Ref. [22]. Copyright (2012) Nature (Color figure online)

polymers. Thus, it was finally concluded that these CP-modified HPGs can be also used as drug delivery vehicle.

The characteristics of an ideal tissue adhesive design should be like—the design must be simple and safe, it must have a tunable setting time depending on the application, it must surely be commercial applicable and the use should be inexpensive, painless, and cosmetic. The conventional cyanoacrylate based tissue adhesives have cytotoxicity where as fibrin based adhesives lack in their adhesion properties. Thus, Wang and co-workers designed a simple and scalable hyperbranched poly(b-amino ester) polymer that can be used as strong wet tissue adhesive [23]. They took Dopamine, an amine-derivative of an amino acid abundantly present in mussel adhesive proteins, and copolymerized via Michael Addition reaction, with a trifunctional vinyl monomer, to form a hyperbranched poly- (dopamine-co-acrylate) (PDA). With fast curing time, low cytotoxicity, and degradable properties, PDA has proved to be a great option as medical sealants and tissue adhesives. The tissue adhesive properties of the PDA polymer was further analyzed by using different curing agents ($FeCl_3$, horseradish peroxidase (HRP)– H_2O_2, fibrinogen and $NaIO_4$) when tested on porcine dermal skin surfaces after curing the adhesive for 15 min, 1 h, and 1 day at room temperature. The test results revealed that Fibrinogen was the best curing agent for achieving a relatively high adhesion strength (37×5.6 kPa) within less time (15 min). Another added advantage with these polymer-based adhesives are due to ester backbone, the poly (b-amino ester)-based polymer was able to degrade under normal physiological conditions via simple hydrolysis mechanism. Further when reinforced with nano-sized HA particles (a basic calcium phosphate $Ca_{10}(PO_4)_6(OH)_2$), PDA can find applications such as tunable bone adhesive for sternal closure (Fig. 7.8) [24]. PDA-based nanocomposite exhibits excellent adhesion and mechanical properties

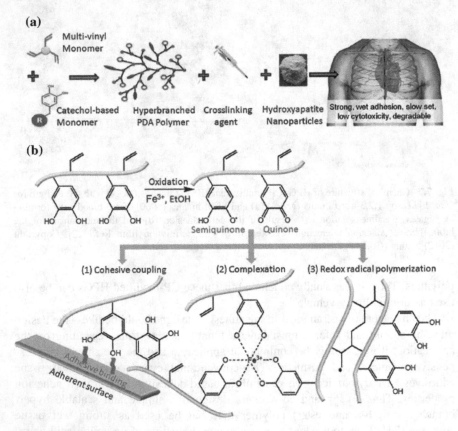

Fig. 7.8 a Strategy of using catechol-modified dendritic PDA polymer nanocomposites for sternal closure and **b** the crosslinking mechanism of PDA with Fe^{3+}. Reproduced with permission from Ref. [24]. Copyright (2014) Royal Society of chemistry

which makes it superior to the conventional existing adhesives. Along with that, PDA adhesives have several other advantages such as easy degradation process which is just inversely related to the healing process and also very low cytotoxicity. Therefore, this PDA-based adhesive could be easily commercialized as cell and tissue adhesives.

7.4 Conclusion

Thus it can be concluded that hyperbranched polymers provide a great matrix for cell adhesion and cell growth. With its multibranching sites and wide range of functionality the hyperbranched polymers are regarded as a great component for tissue scaffolds. On the basis of current researches going on with these special architectural polymers, it can be concluded that the field needs to be explored further for better and more specific target. Hyperbranched polymers have triggered the interest around the researchers to develop more biocompatible and efficient tissue component.

References

1. Lin Y-M, Chrzanowski W, Knowles J, Bishop A, Bismarck A (2010) Adv Eng Mater 12: B101–B112
2. Akdemir ZS, Kayaman-Apohan N, Kahraman MV, Kuruca SE, Gungor A, Karadenizli S (2011) J Biomater Sci Polym Ed 22:857–872
3. Puppi D, Dinucci D, Bartoli C, Mota C, Migone C, Dini F, Barsotti G, Carlucci F, Chiellini F (2011) J Bioact Compat Polym 26:478–492
4. Luo Y-L, Zhang C-H, Xu F, Chen Y-S (2012) Polym Adv Technol 23:551–557
5. Kennedy R, Ul Hassan W, Tochwin A, Zhao T, Dong Y, Wang Q, Tai H, Wang W (2014) Polym Chem 5:1838–1842
6. Zhang JG, Krajden OB, Kainthan RK, Kizhakkedathu JN, Constantinescu I, Brooks DE, Gyongyossy-Issa MIC (2008) Bioconjugate Chem 19:1241–1247
7. Mintzer MA, Grinstaff MW (2011) Chem Soc Rev 40:173–190
8. Yang TF, Chin W, Cherng J, Shau M (2004) Biomacromol 5:1926–1932
9. Das B, Chattopadhyay P, Mandal M, Voit B, Karak N (2013) Macromol Biosci 13:126–139
10. Puppi D, Dinucci D, Bartoli C, Mota C, Migone C, Dini F, Barsotti G, Carlucci F, Chiellini F (2011) J Bioact Compat Polym 26:478–492
11. Xu J, Song J (2010) Proc Natl Acad Sci USA 107:7652–7657
12. Lutz J-F, Weichenhan K, Akdemir O, Hoth A (2007) Macromolecules 40:2503–2508
13. Tai H, Howard D, Takae S, Wang W, Vermonden T, Hennink WE, Stayton PS, Hoffman AS, Endruweit A, Alexander C, Howdle SM, Shakesheff KM (2009) Biomacromolecules 10:2895–2903
14. Hassan W, Dong Y, Wang W (2013) Stem Cell Res Ther 4
15. Nikolovski J, Mooney DJ (2000) Biomaterials 21:2025–2032
16. Burdick JA, Anseth KS (2002) Biomaterials 23:4315–4323
17. Nuttelman CR, Tripodi MC, Anseth KS (2004) J Biomed Mater Res, Part A 68A:773–782

18. Knischka R, Lutz PJ, Sunder A, Frey H (2001) Abstr Pap, Jt Conf—Chem Inst Can Am Chem Soc 221, U438
19. Kainthan RK, Janzen J, Levin E, Devine DV, Brooks DE (2006) Biomacromol 7:703–709
20. Kainthan RK, Janzen J, Kizhakkedathu JN, Devine DV, Brooks DE (2008) Biomaterials 29:1693–1704
21. Oudshoorn MHM, Rissmann R, Bouwstra JA, Hennink WE (2006) Biomaterials 27: 5471–5479
22. Yu X, Liu Z, Janzen J, Chafeeva I, Horte S, Chen W, Kainthan RK, Kizhakkedathu JN, Brooks DE (2012) Nat Mater 11:468–476
23. Zhang H, Bre LP, Zhao T, Zheng Y, Newland B, Wang W (2014) Biomaterials 35:711–719
24. Zhang H, Bre LP, Zhao T, Newland B, Da Mark C, Wang W (2014) J Mater Chem B 2: 4067–4071

Chapter 8
Conclusion

So it is seen that, over the past few years there was a significant progress in the synthesis and development of biocompatible and biodegradable hb polymers for more versatile and sophisticated biomedical uses. Due to tuneable architecture (long and short branches) and functionality, the hb polymers have already shown a great potential of being an "indispensible" in modern biotechnology and biomedical ficlds. In this book we have given best of our efforts to showcase the synthesis strategy and the most latest uses of different hb polymers in the biomedical field. A detailed discussion on the synthesis methodology elucidates relative merits and demerits of different techniques for the synthesis of hb polymers befitting biomedical applications. However, in all the cases, the strategies were useful for the synthesis of hb polymers with tailor made architecture and functionality though. Efforts are still on for development of architectures with even more controlled degree of branching and molecular weight. Owing to unique topological features, abundant functional groups, low toxicity, non-immunogenicity vis-a-vis easy degradation quality has greatly popularized the hb polymers as controlled release matrix, scaffolds for tissue engineering and agents for bioimaging. Clinical trials in many occasion has already been made while some more are in pipeline. The trials were being done on collaborative mode between the laboratory research groups and hospital personnel. On other side, such collaborations can further promote more interdisciplinary research ideas between the interface of biology and polymer chemistry for future innovations in medicinal biotechnology. Few successful applications of hb polymers so far in biological realm has really confirmed its extremely high potential vis-a-vis off shoots a great promise for many more interesting and innovative applications in the years to come.

However, like all earthly materials, it is not a tale of only success. Unsymmetrical structures and broad molecular weight distribution have been the real contenders for hb polymers. These properties were the real culprits debarring hb polymers from many promising applications even in medicinal biotechnology. Since the hb polymers have close analogy to dendrimers, efforts are now being given to wipe off the gap between the analogues so that even the most common and

© Springer Nature Singapore Pte Ltd. 2018
A. Bandyopadhyay et al., *Hyperbranched Polymers for Biomedical Applications*,
Springer Series on Polymer and Composite Materials,
https://doi.org/10.1007/978-981-10-6514-9_8

cheaper monomers can also be utilized to obtain the desired architecture and assembly befitting specific applications. If such effort sees the daylight, will really reduce the production cost and in subsequence will reduce the end product cost as well. In medicinal and therapeutic biotechnology, such an invention would significantly reduce the cost of sophisticated treatments. Thus, this has been the latest research trend along with the development of more specific and stereogenic architectures having a minimal response time. Another challenge is of course is the scale of production of such designer polymers. Although hb polymers in general are easier to scale up but still a steady technology should be developed for these new classes of hb polymers for production in consistent quantity and quality. More target-specific hb polymers are in demand, of late, in medicinal biotechnology specifically to counter mutations of genes of the foreign proteins and bacteria pertaining to various environmental factors. A more complicated and mutative gene requires faster stimuli responsive agents for identification of such foreign agents in a biological system and their subsequent therapy and thus the great challenge is to develop befitted hb architectures capable of similar response from simple monomer systems. We are hopeful such special types of hb polymers and many more similar polymers with more intrinsic modification will be developed in due course of time for more sophisticated yet simple and cheaper therapeutic mode of treatments of the diseased human being.

Printed in the United States
By Bookmasters